Advances in Software Maintenance Management: Technologies and Solutions

Macario Polo
Universidad de Castilla - La Mancha, Spain

Mario Piattini
Escuela Superior de Informatica, Spain

Francisco Ruiz
Escuela Superior de Informatica, Spain

IDEA GROUP PUBLISHING
Hershey • London • Melbourne • Singapore • Beijing

Acquisition Editor:	Mehdi Khosrowpour
Senior Managing Editor:	Jan Travers
Managing Editor:	Amanda Appicello
Development Editor:	Michele Rossi
Copy Editor:	Jane Conley
Typesetter:	Tamara Gillis
Cover Design:	Integrated Book Technology
Printed at:	Integrated Book Technology

Published in the United States of America by
Idea Group Publishing (an imprint of Idea Group Inc.)
701 E. Chocolate Avenue, Suite 200
Hershey, PA 17033-1240
Tel: 717-533-8845
Fax: 717-533-8661
E-mail: cust@idea-group.com
Web site: http://www.idea-group.com

and in the United Kingdom by
Idea Group Publishing (an imprint of Idea Group Inc.)
3 Henrietta Street
Covent Garden
London, WC2E 8LU
Tel: 44 20 7240 0856
Fax: 44 20 7379 3313
Web site: http://www.eurospan.co.uk

Library of Congress Cataloging-in-Publication Data

Polo, Macario, 1971-
Advances in software maintenance management : technologies and solutions / Macario Polo, Mario Piattini and Francisco Ruiz.
p. cm.
ISBN 1-59140-047-3 (hardcover) — ISBN 1-59140-085-6 (ebook)
1. Software maintenance. 2. Computer software—Development—Management. I. Piattini, Mario, 1966- II. Ruiz, Francisco, 1967- III. Title.
QA76.76.S64 P64 2003
005.1'6—dc21

2002014187

British Cataloguing in Publication Data
A Cataloguing in Publication record for this book is available from the British Library.

Advances in Software Maintenance Management: Technologies and Solutions

Table of Contents

Preface

> It is a good thing to practice some bad habits such as smoking, eating pork, drinking over your limit or not doing any physical exercise, so that if one day you fall ill, your doctor, to make recover, will have something to ban. But if you are all virtue, you will have no further room for improvement and falling ill will take you to your death bed.
>
> Luis Landero, in *Games of the Late Age*.

The choice of the quotation that starts this preface has not been casual. In fact, software in execution suffers many bad habits that, fortunately for software services companies, produce more and more work every year. From the point of view of software maintenance, such imperfections have their origin in the software itself, when some defects must be removed; in users, when they ask for new functionalities to be added; and in the changing technological environment, when the software must adapt to a new environment. On the other side, and mapping Lehman Laws (Lehman, 1980) with the last sentence of the quotation, non-changing software is non-used, dead software.

According to the ISO/IEC (1995) terminology, the software maintenance process is activated "when the software product undergoes modifications to code and associated documentation due to a problem or the need for improvement or adaptation." In spite of this definition, which is very similar to that of ANSI-IEEE (1990), the ignorance of maintenance activities may lead to underestimating its importance, since there is a tendency to associate software maintenance only with corrective activities. However, several authors (McKee, 1984; Frazer, 1992; Basili et al., 1996; Polo, Piattiani, & Ruiz, 2001) have shown that perfective interventions receive the most effort of maintenance.

From the seventies, software maintenance is the most costly stage of the software life cycle (see Table 1), and there are no reasons to think that the situation will change, since novel environments and technologies require great maintenance efforts to keep software products in operation. For Brereton, Budgen, and Hamilton (1999), maintenance of hypertext documents will become a serious problem that requires immediate action, since they share many characteristics (structure, development process, economical value) with classical software products. According to Lear (2000), many legacy applications written in COBOL are being adapted to be integrated with current technologies, such as e-commerce.

There are organizations that devote almost all their resources to maintenance, which impedes new development. Moreover, maintenance necessities increase as more software is produced (Hanna, 1993), and its production has al-

Table 1. Evolution of maintenance costs.

Reference	Year	% Maintenance
Lientz and Swanson (1980)	1976	60%
Pigoski, (1996)	1980-1984	55%
Schach (1990)	1987	67%
Pigoski (1996)	1985-1989	75%
Frazer (1992)	1990	80%
Pigoski (1996)	90's (prev.)	90%

ways shown a growing tendency. On the other side, big programs never are complete, but are always in evolution (Lehman, 1980). Ramil, Lehman, and Sandler (2001) confirm this old theory 21 years later.

PROBLEM CAUSES

In spite of this, software organizations still pay more attention to software development than to maintenance. In fact, most techniques, methods, and methodologies are devoted to the development of new software products, setting aside the maintenance of legacy ones. This problem is also common among programmers, for whom maintenance is "less creative" than development; in fact, many legacy systems use old and boring programming environments, file systems, etc., whereas programmers prefer working with new, powerful visual environments. However, the same software evolves and must continue to evolve along years and, to their regret, programmers devote 61% of their professional life to maintenance, and only 39% to new development (Singer, 1998).

The lack of methodologies may be due to the lack of a definition of the software maintenance process. For Basili et al. (1996), the proposal and validation of new methodologies that take into account maintenance characteristics are a must. Also, Pigoski (1996) says that there is little literature regarding maintenance organizations.

PROPOSED SOLUTIONS

It is clear that maintenance organizations require methodologies and techniques that facilitate software maintenance, decreasing costs and difficulties. There exist different types of partial solutions for software maintenance. Depending on their nature, they can be classified into:

- Technical solutions, that assist in certain moments of maintenance interventions. Reengineering, reverse-engineering or restructuration techniques are some examples.

- Management solutions, that are mainly based on quality assurance, structured management, change documentation and use of specialized human resources.

ORGANIZATION OF THE BOOK

This book collects proposals from some of the best researchers and practitioners in software maintenance, with the goal of exposing recent techniques and methods for helping in software maintenance. The chapters in this book are intended to be useful to a wide audience: project managers and programmers, IT auditors, consultants, as well as professors and students of Software Engineering, where software maintenance is a mandatory matter for study according to the most known manuals—the SWEBOK or the ACM Computer Curricula.

Chapter I, by Ned Chapin, sets a foundation for software maintenance management, analyzing the influence of business rules on maintenance of information systems. Mira Kajko-Mattsson presents in Chapter II the main problems related to the corrective maintenance, and she presents some management solutions for them, strongly supported by a solid theoretical and practical basis.

In Chapter III, Fabrizio Fioravanti analyzes the impact of the recent programming approach of XP on software maintenance, explaining the relationships of eXtreme Programming with maintenance projects. Then, in Chapter IV, Perdita Stevens introduces the idea of using design and process patterns in software maintenance; she also emphasizes the convenience of organizations writing their own set of specific patterns.

William C. Chu, Chih-Hung Chang, Chih-Wei Lu, Hongji Yang, Hewijin Christine Jiau, Yeh-Ching Chung, and Bing Qiao devote Chapter V to maintainability, the desirable property of software systems that would make maintenance work easier. They study the influence of the use and integration of standards on maintainability.

In Chapter VI, Lerina Aversano, Gerardo Canfora, and Andrea De Lucia present a strategy for migrating legacy systems to the Web. After presenting their method, the authors explain an application experience.

In Chapter VII, Norman F. Schneidewind analyzes the relationship between requirements risks and maintainability. It illustrates this study with some empirical results of a number of big spatial projects.

Harry M. Sneed gives us an excellent lesson on software maintenance cost estimation in Chapter VIII, presenting a clear, systematic method for this task.

Macario Polo, Mario Piattini, and Francisco Ruiz devote Chapter IX to present Mantema, a methodology for software maintenance with outsourcing support and explicit definition of process models for different types of maintenance. Mantema is integrated into the MANTIS environment, that is presented in Chapter X by these very same authors and Félix García.

REFERENCES

ANSI-IEEE (1990). ANSI/IEEE Standard 610: IEEE standard glossary of software engineering terminology. New York: The Institute of Electrical and Electronics Engineers, Inc.

Basili, V., Briand, L., Condon, S., Kim, Y., Melo, W. & Valett, J.D. (1996). Understanding and predicting the process of software maintenance releases. In *Proceedings of the International Conference on Software Engineering,* (pp. 464-474). Los Alamitos, CA: IEEE Computer Society.

Brereton, P., Budgen, D. & Hamilton, G. (1999). Hypertext: The next maintenance mountain. *Computer, 31*(12), 49-55.

Frazer, A. (1992). Reverse engineering-hype, hope or here? In P.A.V. Hall (Ed.), *Software Reuse and Reverse Engineering in Practice* (pp. 209-243) Chapman & Hall.

Hanna, M. (1993, April). Maintenance burden begging for a remedy. *Datamation,* 53-63.

ISO/IEC (1995). International Standard Organization/International Electrotechnical Commission. *ISO/IEC 12207: Information Technology-Software Life Cycle Processes*. Geneve, Switzerland.

Lear, A.C. (2000). Cobol programmers could be key to new IT. *Computer, 33*(4), 19.

Lehman, M. M. (1980). *Programs, life cycles and laws of software evolution. Proceedings of the IEEE, 68*(9), 1060-1076.

Lientz, B.P. & Swanson, E.F. (1980). *Software Maintenance Management.* Reading, MA: Addison Wesley.

McKee, J.R. (1984). Maintenance as a function of design. In P*roceedings of AFIPS National Computer Conference in Las Vegas*, 187-93.

Pigoski, T. M. (1996). *Practical Software Maintenance. Best Practices for Managing Your Investment*. New York: John Wiley & Sons.

Polo, M., Piattini, M. & Ruiz, F. (2001). Using code metrics to predict maintenance of legacy programs: A case study. *Proceedings of the International Conference on Software Maintenance.* Los Alamitos, CA: IEEE Computer Society.

Ramil, J.F., Lehman, M.M. & Sandler, U. (2001). An approach to modelling long-term growth trends in large software systems. *Proceedings of the International Conference on Software Maintenance*. Los Alamitos, CA: IEEE Computer Society.

Schach, S.R. (1990). *Software Engineering*. Boston, MA: Irwin & Aksen.

Singer, J. (1998). Practices of software maintenance. In Khoshgoftaar & Bennet (Eds.), *Proceedings of the International Conference on Software Maintenance,* (pp. 139-145) Los Alamitos, CA: IEEE Computer Society.

ACKNOWLEDGMENTS

We would like to thank all the authors for their excellent contributions. We want also to acknowledge the guidance of Michele Rossi, our Development Editor, for her motivation and patience, and the continuous support of the Idea Group Publishing staff for this project.

Macario Polo, Mario Piattini, and Francisco Ruiz
Ciudad Real, Spain, September 2002.

Chapter I

Software Maintenance and Organizational Health and Fitness

Ned Chapin
InfoSci Inc., USA

This chapter sets out a foundation for the management of software maintenance. The foundation highlights a fundamental for managing software maintenance effectively—recognition of the connection between managing change in organizations and the organizations' health and fitness. The changes arising as a consequence of software maintenance nearly always impair or improve that health and fitness. The basis for that impact lies in how business rules contribute to the performance of systems, and in how systems contribute to the performance of organizations. An understanding and application of how changes from software maintenance on business rules and systems affect the performance of organizations provides a basis for management action. With that added confidence, managers can reduce impairing their organizations' health and fitness and can increase positive action to improve their organizations' health and fitness via software maintenance.

INTRODUCTION

Software maintenance provides a vital path for management in preserving and building organizational health and fitness. Since software maintenance also is the usual means for implementing software evolution, that vital path takes on additional significance for management. As explained later in this chapter, that vital path provided by software maintenance is a main way of making intentional changes in how organizations work. In the contexts in which they operate, how organizations work is manifested in their organizational health and fitness.[1]

To manage these intentional change processes effectively, managers need to understand eight activities and attributes of organizations and appreciate and apply their characteristics. Managers need to know the roles of systems and how the components of systems interact in getting work done in organizations. Since systems typically implement organizations' business rules, managers can introduce and guide change in their organizations by giving emphasis to the choice and relevance of the implemented business rules.

For these reasons, this chapter starts out by giving attention in its first main section to depicting the lay of the land with respect to the activities and attributes of organizations. The second main section is devoted to the role of systems in organizations, and the third to business rules. The fourth main section delves into systems and the operational performance of organizations and relates them to organizational health and fitness. The fifth main section tells about the impact of changes and the processes of changing organizations, systems, and business rules. Those five main sections support the sixth main section, which covers ways that software maintenance and software evolution contribute to organizational health and fitness. Two sets of ten activities, one negative set and one positive set, are summarized in the seventh main section, on the potential of software maintenance for determining organizational health and fitness. After noting maintenance maturity, this chapter then ends with a conclusion section.

THE LAY OF THE LAND

For the purposes of discourse here, let us limit the term ***organization*** to entities deliberately created and operated by human beings for human-designated purposes (Drucker, 1973a). Thus, the astronomical organization commonly known as the solar system is not considered here—it was not created by people for their purposes. And the human digestive tract with its major components including salivary glands, esophagus, stomach, and intestines is not here considered as an organization—although human couples may act deliberately to create children, the digestive tract is part of the child, and cannot currently be separately created and be viable. Some examples of organizations as considered here in this chapter are: a government agency to collect value-added tax (VAT); a dictation and word-processing pool serving a group of barristers; a multinational, globally operating, vertically integrated producer-distributor of petroleum-based products; a fútbol (soccer) club; a coal mining corporation; a fishing fly (bait) provider operated by one person working part-time; a temporary emergency flood relief charity; and a public accounting service.

Figure 1 on the next page diagrams an organization in operational terms. Most organizations are composed of sub-organizations. Hence, when human beings (people or personnel) are members or components of an organization, an individual person may be and usually is a member of or a component in more than one suborganization.

Although no analogy is perfect, an organization can be regarded as if it were a natural living biological entity. Eight activities and attributes of organizations are also activities and attributes of living biological entities:

- They *enclose self within boundaries* that separate the self from its environment and milieu.[2] For biological entities, the boundaries may take such physical forms as membrane, hide, cell wall, bark, skin, scales, etc. In addition, some animal entities mark territorial boundaries, as by scents, scratch marks, etc. In organizations, the boundary may take both physical forms such as a building's walls, or territorial forms such as signs, different cultural practices, different ways of doing things (such as the Navy way, the Army way, etc.).
- They *preserve boundary integrity* to reduce breaches of the boundaries. For biological entities, this may take behavioral forms such as posturing, grooming, threatening, hardening of the boundary material, refreshing markings, etc. Organizations use many ways to try to preserve their boundary integrity, such as advertising, using spokespersons, holding meetings, seeking endorsements, screening applicants, etc.
- They *acquire materials* from the environment to incorporate within the self. These are inward-flowing transports across the boundaries. For a biological entity, the materials are its diet and its sources of energy, such as sunlight. Organizations take in materials (such as money, electricity, iron ore, cardboard, etc.) both from the environment and from its milieu such as from other

Figure 1: An operational view of an organization

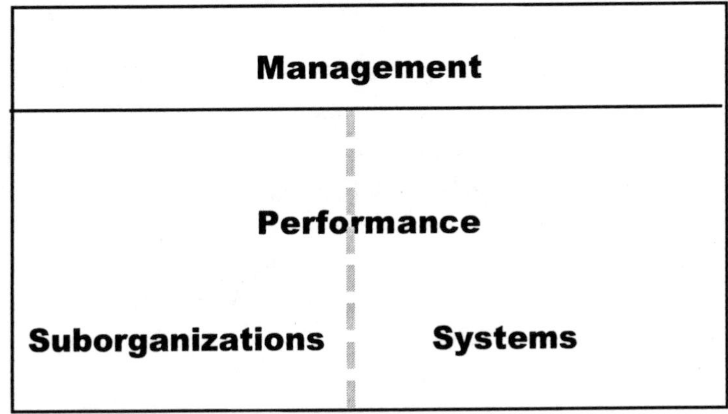

organizations. For example, an organization may acquire money from customers, electricity from a power generating organization, iron ore from a mining organization, cardboard from a distributor, etc.

- They *use acquired materials* to grow and replenish the self. This is a negative entropy activity keeping order in the self and nearly always requires an expenditure of energy. For biological entities, the acquired materials are selectively used as the energy and the substances to grow or replenish the self and for the self to reproduce more of its own specific kind, all with some release of energy—often as chemical changes, motion or heat. Organizations use the acquired materials (such as money, electricity, iron ore, cardboard, etc.) to grow or replenish the self, all with some release of energy—usually as motion or heat. Organizations differ from biological entities in that organizations rarely use the acquired materials to reproduce more of their own specific kind; instead, organizations usually generate products or services.
- They *purge non-self* from within the boundaries. These are outward-flowing transports across the boundaries. For biological entities, the purging usually is of generated waste and of unwanted materials that have come to or inward across the boundary, such as poisons, infections or parasites. For some biological entities, the purging may be part of the reproduction process. Organizations use many ways to purge non-self from self, such as shipping of product, staff training, internal disciplinary action, promotion, local standard operating practices (SOP) enforcement, personnel discharge, competitive analyses, etc.
- They *respond to external stimuli* in an ongoing continuous and concurrent manner about the interaction of the entity with its environment and milieu. The links between the stimuli and the responses usually are complex and varied. For biological entities, the stimuli may give rise to behavioral and state changes as responses, such as wilting, running, eating, defending, blossoming, etc. Unlike biological entities with their built-in stimuli sensors and linkages to built-in response capabilities, organizations have no inherent links between the stimuli and responses. Instead, organizations have to select and actualize sensory input handling, output actions and states, and the links between them.
- They *respond to internal stimuli* in an ongoing continuous and concurrent manner about what goes on within the self. The links between the stimuli and the responses usually are complex and varied. For biological entities, an example is gas exchange processes that occur at the same time that fluid transport and internal fluid circulation occur, and how changes in one may affect the others. Because of the typical presence of interacting people in an organization, most organizations have a lot of concurrent activities going on (customer relations, custodial services, production planning, invoicing, etc.),

with the people often providing much of the sensory apparatus. But the links to the typically wide selection of responses tend to be inconsistent and erratic, because of relying on people to provide or do the linkages. For example, because of the voluntary sudden departure of an employee, the other employees have to change what they do and when and how they do it, to pick up on-the-fly the workload change caused by the departure until a replacement is hired and broken in, if the organization's operational performance effectiveness is to be preserved.

- They *avoid illness and infirmity* to the extent they do the seven activities listed above. Some degree of success is needed in all seven for the entity to have good health and fitness, although partial substitutions are sometimes effective. For biological entities, animals and insects have more opportunity to act to improve their health and fitness than do plants, because plants typically have much less mobility. Organizations have the most opportunity to improve their health and fitness—i.e., their operational performance—as explained later in this chapter.

Organizations normally enjoy one attribute that biological entities do not have—an ability to shape-shift and function-shift at will. Biological entities do shape-shift and function-shift, but in a set sequence and on an inflexible schedule. For example, a tomato plant make take the shape of a tomato seed, of a tomato seedling, and as a fruiting tomato plant, but only in that sequence and only among those forms and their associated functions. Vertebrate animals take the shape of a gamate (egg or sperm), fetus, youngster and adult, but only in that sequence and only among those forms and their associated functions. Many bacteria have two forms and two sequences, cell to two separate cells, or cell to spore to cell, but only in those sequences and only among those forms and their associated functions.

An organization, by contrast, has much more latitude to shape-shift and function-shift, within limits as noted later. An organization can select its own constituent components and how they are to interact, and change both at any time. An organization can change its internal processes at any time. An organization can choose the form, substance, and position of its boundaries, and change them at any time. An organization can select the products or services it produces and distributes, and how and when it produces and distributes them. An organization can select what means of self-protection to deploy, and how and when it deploys and engages them at any time. And an organization can determine what it will treat as stimuli and the linkages to select what responses to make, when, and how, and can change them at any time.

The common limitation on an organization's ability to shape-shift and function-shift arise from two main sources: access to materials (resources), and management decision-making. The usual resource limitations are prices, capital, personnel skills complement, implementations of process/operations engineering results, and feedstock availability and quality. The usual management decision-making limitations are narrow management vision, restrictions in the stimuli, unrecognized opportunities in linking stimuli and responses or response availability, and unrecognized opportunities in either the milieu or the environment, or in the use of materials (resources).

THE ROLE OF SYSTEMS

In an organization, a ***system*** is a combination for a human-selected purpose within a boundary of the services of personnel, machines, materials, and methods.[3] The services only are active because the personnel, machines, materials, and methods are not destroyed or consumed (like the blackness of ink on the whiteness of paper) as the system operates, except for some materials (such as electricity). Figure 2 diagrams a system in terms of services, while Figure 3 diagrams a system in operational terms, and Figure 4 diagrams a system in terms of interactions and components. Systems are often composed of subsystems. Systems usually are nested and, when so, share some boundaries. The boundary of a system or subsystem usually is identified in terms of what it takes in from its environment or milieu, and what it puts into its environment or milieu. For our focus here on information systems, the boundary is usually identified by the data accepted as input and the data provided as output.

An organization uses systems to provide most of an its operational performance capability in three main ways:

Figure 2: A services view of a system

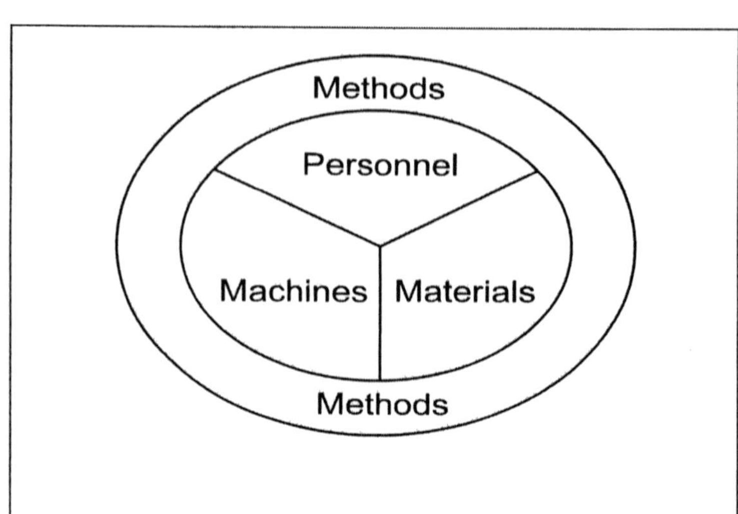

- The most common way is to use interacting systems to accomplish the creation and distribution of products and services along with the associated overhead. For example, to produce electric motors, the production department uses a full enterprise resource planning (ERP) system or a combination of inventory systems (raw material, work in process, and finished goods), bills of materials system, production control system, logistics system, quality control system, plant and equipment support system, purchasing and supply chain system, and cost accounting system. Stimulus-responses links with feedback are common, and the timing of action is normally important in such systems; for information systems, a common implementation form is online transaction processing (OLTP).
- A second way is to use interacting systems (primarily information systems) to implement in the organization the equivalent of a (biological) nervous system for providing responses to stimuli other than those directly involved in the product/service systems. These more general systems deal with data relevant for line and staff functions in an organization, such as building repair status, accounts receivable, human resources, advertising, and site security. Such systems involve a coordinated use of data acquisition, data reduction, data storage (in information systems, databases and files), data analysis, data processing, and data reporting to provide data on a periodic or an as needed basis to destinations within the organization (such as personnel and machines) and without (such as external stakeholders, regulatory agencies, etc.).
- A third and less common way is to use specialized computerized information systems for helping management in policy determination and planning for the

Figure 3: An operational view of a system

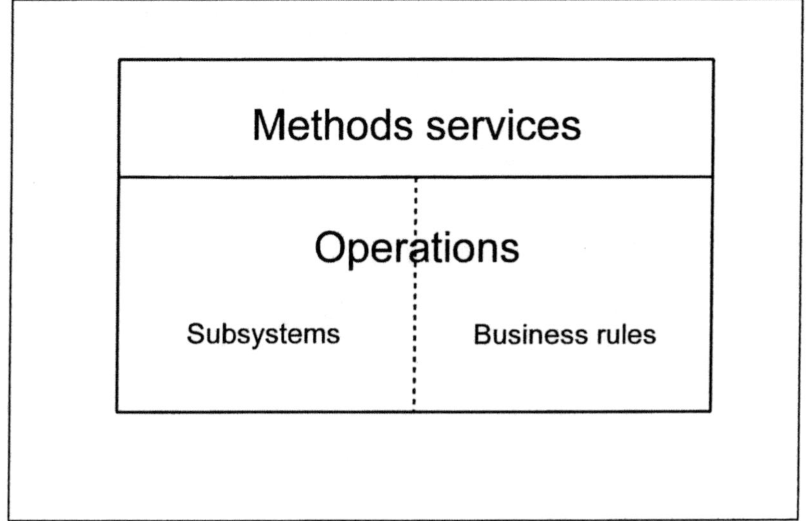

future. These systems usually comprise an organization's complement of "what if" systems for its decision support systems (DSS) and executive information systems (EIS).

In understanding the role of systems in organizations, managers who have stayed with one organization for many years tend to have an advantage. This is especially true if they have held a variety of positions and have had promotions into positions at increasingly higher levels of management. The personnel working in organizations face a diversity of situations, but differ by the place (suborganization) where they work, where that is, such as accounts receivable, scrap recycling, parking control, new product development, etc. Also, many of the situations faced are recurrent, some very frequently (such as receiving a full payment on an account receivable), some are occasional (such as receiving a payment applying to an invoice that has not yet been submitted), and some are rare (such as receiving a full recovery with interest on a closed written-off bad debt account). In addition, some situations are non-recurrent, such as disaster recovery from a tornado strike.

For the non-recurrent situations, managers may make plans for the eventuality that the situation might come to pass. But some situations come unanticipated, and managers typically respond to them with improvising an ad hoc response for the personnel in the organization to do in order to handle the situation. In handling these, existing systems may be pressed into service for whatever contribution they may be able to make.

Handling the recurrent situations is the normal province of systems, especially of computerized information systems. As emphasized later in this chapter, handling

Figure 4: A view of an information system in terms of data flows and subsystems

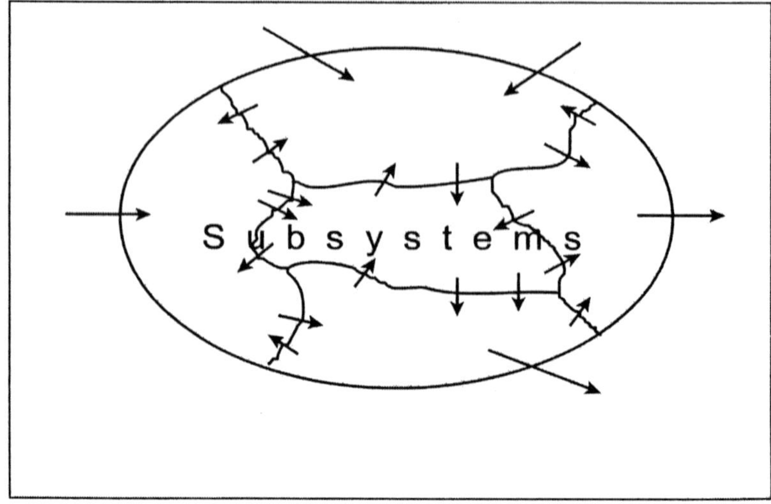

recurrent situations by means of using information systems gives consistency to the organization's performance, and usually also reliability, economy, and appropriate timing. For example, consider the handling of mailed-in payments on accounts receivable. After a machine has opened the mail, personnel examine the documents to verify or correct the account designation and the amount received. The actual payments are then routed to a payment processing system, while the account and amount received data are machine read with unreadables manually entered. The computer (usually via an online transaction processing system) validates each transaction, rejecting invalids and applying valids to the correct accounts to update them in the organization's database or file of customer records. The computer also updates running summaries and provides data for associated systems, such as for market analysis, auditing, cash reconciliation and cash management. Personnel examine the invalids and after making inquiries of the database or file, correct the transaction data and reinsert the transaction data into the transaction stream. The brief high-level summary just given is generic; in practice, each organization uses a customized system that does the specifics of the operations in the systems a little differently.

THE ROLE OF BUSINESS RULES

An organization uses systems to implement or embody within that system's boundaries and as part of its operations, that organization's "business rules." A ***business rule*** determines an organization's behavior in recurrent situations. A business rule includes a ***decision portion*** concerned with detecting or ascertaining the situation (but normally without formally identifying the name of the situation), and also one or more ***action portions*** specifying or directing what is to be done and how it is to be done in the specific situation. An action portion may include nested business sub-rules. Figure 5 diagrams a business rule in operational terms.

Business rules may be carried out by many kinds of systems. For example, a business rule in an automatic machine tool used for making gears intended to become part of an automobile automatic transmission, may specify the cutting of a gear tooth with a specific machining tolerance on a certain type of gear blank cast using a particular alloy. Or, for example, a business rule may specify to a person doing sales work how much may be spent on food and beverages when entertaining a major repeat customer's representative.

In information systems, the decision portion of a business rule typically is represented as a predicate, such as by the equivalent of a set of nested IF statements evaluating the current values of specified control variables. A simple example of a decision portion (expressed in pseudocode or semicode) is IF (SCRAP-RATE > NORMAL-SCRAP-RATE) AND (TOOL-AGE < EXPECTED-TOOL-LIFE). An action portion of a business rule in an information system typically is represented as a series of imperatives that assign values to or change the current values of processing variables or of control variables, or some of both. A simple example of an action portion (expressed in pseudocode or semicode) is SET STATUS-FLAG TO 6; INCREMENT STEP-COUNTER BY +3; MOVE CURRENT-CLIENT TO MANUAL-REVIEW-LIST.

As noted earlier, in most systems and in nearly all information systems, the business rules form a nested set to provide part of a system's performance capability. The rules within a nested set form either a hierarchy or a network of business rules. In a nested set, any particular business rule is active (fired) only when the values of each decision portion are appropriately satisfied. Figure 6 provides a very simple example of an hierarchical nesting of business rules. In it, Business Rule #2 is nested within Business Rule #1. For Business Rule #2 to become active, the value of Control Variable A in Business Rule #1 must be greater than the value of Control Variable B at the time that the decision portion of Rule #1 is started. Note, however, that either ACTION G or ACTION H or the action portion of any subsequently active business rule might later change the value of any of the A or B or C or D control variables. Any such subsequent changes in those control variables does not terminate, invalidate, or reverse the consequences of all previously initiated action portions of business rules, such as of Rules #1 and #2.

Especially in large information systems, the number of business rules' decision portions that must be satisfied for any specific business rule's action portion to become active may be any positive integer depending upon the position of the specific rule in the system's network or hierarchy of business rules. In addition, a system may be called for execution by the action portion of a business rule located

Figure 5: An operational view of a business rule

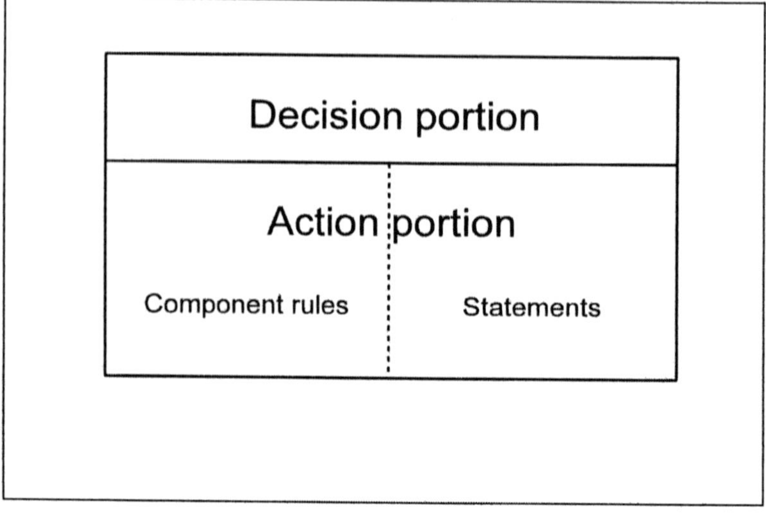

anywhere in the network or hierarchy of business rules of some other system. Furthermore, the possible execution pattern for a set of systems may be a network or a hierarchy, and the number of systems in the set of systems may be one or any positive integer.

In most systems and in nearly all information systems, only one necessary constraint applies to the business rules. That constraint is that values for the rule's control variables must be available at the start of the execution of the rule's decision portion. A value may come directly from a senor, or indirectly from an A-to-D (analog-to-digital) converter, or from previously saved (stored) sources such as a cache or file or database, or from just produced data from the system, or from parameter passing, or more rarely from other sources. As long as this constraint is met, any business rule is a candidate for inclusion in any information system. However, a consequence of the activity attributes of organizations reviewed earlier is that almost all information systems in practice meet a relevance constraint—human-created information systems are deliberately made only with business rules that either the organization's management or system personnel believe are relevant to that system's human-set purpose. Some of the major consequences of this human-dependent relevance constraint are covered next.

SYSTEMS AND THE OPERATIONAL PERFORMANCE OF ORGANIZATIONS

Just as systems are largely built of business rules for the operational performance the rules provide, so also organizations are largely built of systems in order

Figure 6: A simple example of nested business rules, represented in three ways: Chapin chart (Chapin, 1974); ANSI flow diagram (Chapin, 1971); and UML activity diagram (Rumbaugh, Jacobson, & Booch, 1999)

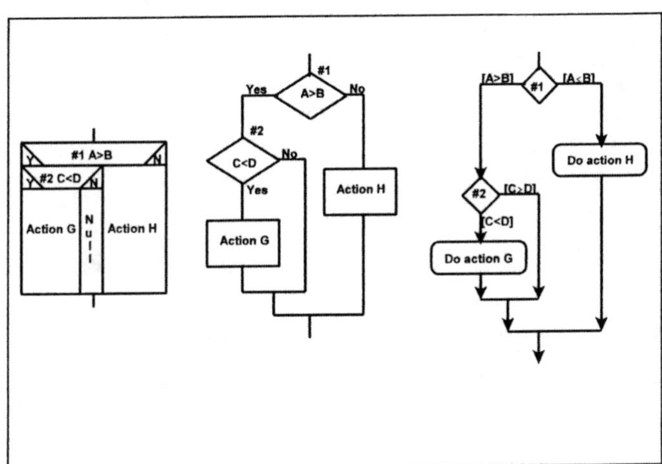

to provide the operational performance of the organizations. By using business rules integrated into sets of systems, the organization can provide concurrent and consistent performance for recurrent (repetitive) situations, each time and however they occur. Non-recurrent situations rarely are the focus of systems because of the relatively lower usual cost of having an employee or contractor create and execute an ad hoc performance, attempting to meet the organization's need in a way considered by management to be appropriately beneficial to the organization. Since most situations that an organization experiences are recurring situations, management normally "delegates" the handling of almost all recurring situations to specific systems and enforces that delegation. The delegated situations run the gamut from the tiny, minor, and trivial to the enormous, major, and critical.

For a low-level example, does a person telephoning four random times into a small women's ready-to-wear shop named "Alice's" hear each time as an initial greeting "Alice's has it. What would you like?" Or does the caller telephoning four random times hear on each call respectively as the initial greeting "Don't bother us—go call someplace else" or "Alice's has what you need, so please come in and get it" or "Wrong number!" or "Lithe Shops gave me better service than Alice's. Do you really want to talk to someone here?"

For a high-level example, do an organization's quarterly financial reports all use the same "generally accepted accounting principles" in a consistent manner? Do they all use the prior report's data as the base for comparison and report the effects of similar transaction is a consistent way? Or, for instance, do the 2nd and 4th quarterly reports value the cost of turnover (sales) using a first in, first out (FIFO) inventory, but for the 1st quarterly report use last in, first out (LIFO), and for the 3rd quarterly report use LIFO for some products and FIFO for others? And is standard costing used for all products, except for the 4th quarterly report which uses "estimated costs" for the six highest volume products?

As noted previously, systems combine services from four sources. Commonly all four services are active. For example, in the current system used in major seaports in loading or unloading containerized seagoing ships, all four are present and active—personnel, machines, materials, and methods. And in an automated telephone answering system, all four services are present and active. But in some situations, one of the services may not be involved, with the following usual consequences:

- The absence of the methods service yields inconsistent performance from the system.
- The absence of the machines service makes people do most of the system performance themselves, for example, in a wine tasting system.
- The absence of the materials service is rare and when present shifts work to the personnel and the machines, for example, causing system reliance on

human memory instead of paper records or computer storage, and contributing to human on-the-fly direction of system performance, as in a solo piano improvisation music delivery system.

- The absence of the personnel service is very rare, for it is a characteristic of a completely automated system where human beings have delegated their usual roles mostly to the methods service, with the rest delegated to the machines and materials services.

While the four services are partially substitutable for each other, substitutions nearly always change the operational performance of the system—they may improve it or degrade it.

Because of the tight connection between the operational performance of a system and the operational performance of the organization, the management of an organization usually specifies the intended purpose of a system and of the operational performance expected of it. Implementing those brings into play the work of the systems personnel who typically try to make a practical and effective actualization or instantiation or implementation of management's expressed specification. For example, the systems personnel might determine that for the system to be operationally effective, it will have to have access to and be supported by two databases, neither of which was part of management's specification and neither of which exist currently.

Also, although usually not stated by management in its specification for a system, the structure or architecture of a system is normally left to the systems personnel. The structure of a system usually is best when it models the structure of the situation in which it is to perform. While this is commonly recognized by systems personnel, especially for real-time systems, it currently is rarely recognized by management. That is, a system or set of systems usually models the management desired or implied fit of the organization in its milieu and environment. The systems are like psychological projections of management's vision, attempting both to reshape the environment and to shape-shift and function-shift the organization to be operationally effective in the environment and milieu management perceives.

THE IMPACT OF CHANGES

Changing an Organization

At any time, management can change its vision of the organization's fit and of the environment and milieu. Management can create new organizations, or terminate existing organizations. Management can change the boundaries and the mission, task, or function of an organization at will, as well as how management wants the organization to perform. In changing the boundaries of an organization, management can also rearrange, reconfigure and reconstitute organizations. It may divide an organization into component organizations, or compose an organization from existing organizations, creating hierarchies or networks of organizations, thus making the prior existing organization into component sub-organizations. Nearly all but the very smallest organizations have or consist of component sub-organizations. Here in this Chapter, the unqualified term "organization" refers to any organization in any position and with any human-designated purpose.

To make any created or modified organization viable, management must tacitly or explicitly link the organization to a purpose and enforce its realization through the organization's operational performance. To give an organization an operational performance capability, management has to specify or to accept, either explicitly or by default, the organizational and system boundaries and the systems the organization is to use. Management may specify shifting those boundaries or adding to, modifying, or terminating these systems at any time.

Changing a System

The same management action that specifies a system can also specify a change in a system. A management-specified change in a system has the effect of making a change in a system's operational performance. Furthermore, such a change in a system's operational performance also results in a change in the organization's operational performance. Such management-specified changes are typically based on the external (management visible) characteristics of a system. Those are external with respect to the specific system, not necessarily external to the organization. Some of the boundaries of a system may coincide with some of the boundaries of an organization, as when an information system used by an organization accepts as a stimulus something from another organization in its milieu or from the organization's environment. Otherwise, the boundaries of a system normally coincide with parts of the boundaries of some of the organization's other systems where those boundaries are internal to the organization's boundaries.

A management-specified change in a system often necessitates, in the judgment of systems personnel, other changes that support or facilitate it. An information system example is when management specifies that products are to be shown hereafter in a value of turnover (sales) sequence in an existing report. The systems personnel recognize that this will require a separate sort processing for each territory, something missing from the management specification. Hence the systems personnel generate consequential changes to make the sorted data available in order to implement the management-specified change in report content and format.

Because the changes, whether direct (management-specified) or consequential, affect a system, any or all of the four services in system may be affected. A change in one service usually ripples to make one existing other service be no longer a good fit. If the fit is left impaired, then the organization's operational performance is also impaired. For example, assume that management requires that the currently manually applied rules to assess customer credit worthiness are to be replaced with a different set of rules effective over the weekend. Furthermore, management requires that the new rules are to be applied by a computer at the time of each transaction with a customer, that equivocal cases are to be flagged for personnel review, and that the review results are to be given to the computer before the transaction is allowed to be completed. The new requirement changes personnel roles and shifts the skills repertoire the personnel need. It also changes the timing, both for the personnel services and for the computer services, and introduces a new dependency between the computer and personnel services. How are both the computer and personnel business rule changes to be accomplished over the weekend to make the required system change without impairing the operational performance of the organization?

Changing a Business Rule

Management-specified changes in systems are normally attempts by management to improve the operational performance of the organization. In practice, some of these management-specified changes have the effect either of shape-shifting or function-shifting the organization, or both, in terms of the interaction of the organization with its environment and milieu. The consequential changes from the systems personnel normally attempt to contribute to or make possible the carrying out of the management-specified changes.

To accomplish change in the systems, the systems personnel address the roles of the business rules in the systems, whether they involve the personnel, the machines, the materials, or the methods, or any combination of them. In doing this, the systems personnel must confirm (and hence leave unchanged) or modify four aspects of the business rules in the system: the stimuli, the responses, the network linking the stimuli with the responses, and the timing of the responses. Existing boundaries and nesting among the rules and systems (and sometimes the organization too) complicate the work of the systems personnel.

In a business rule, the stimuli act as the control variables and as the incoming items for the action portion of the rule. The responses are the outgoing results from the selected action portion of the business rule. Together as noted earlier, the stimuli and the possible responses mark the boundaries of the business rule. The linking network is the algorithm or heuristic within the rule that uses the control variables to trigger the selection of the response in the action portion of the rule. The timing may be a default timing determined by the machines or materials, or be a variable timing determined on-the-fly by the personnel in the system, or be a chosen timing determined by the method implemented in the action portion of the business rule.

Restrictions are from the specified choice made for the services. Stimuli may be limited or untimely. For example, people are blind in the ultraviolet and microwave frequencies, but some sensor equipment is sensitive to frequencies that people are not. Very importantly, specifications by management or by the systems personnel may result in blindness in business rules. An example in a financial forecasting information system is to fail to include the trends in prices for items normally purchased by the organization, or to fail to use current trend data instead of obsolete data. Another example is a decision, deliberately or by default, to exclude from a market forecasting system the prices of products from competitor organizations, or to exclude competitor updated actual prices and use instead prices as if they were constants.

Responses may also be limited or untimely. For example, a response from a business rule in an information system may be to decrease immediately by 20% the flow of sulfuric acid in an industrial process. But, if the response is implemented by a person more than two days after the system's business rule directed the flow reduction to occur, spoilage or other damage to the organization's assets may have been occurring. Alternatively, the business rule could have provided the response to a digital to analog (D-to-A) converter attached to the flow control valve on the acid supply to effect an immediate flow reduction, assuming such a D-to-A converter and associated valve were among the machines in the system.

The linking network may be incompetent or act in an untimely manner. For example, irrespective of the values of the control variables, the linking network may always select the same response ("stuck on one"). Or the linking network may operate too slowly, as in an information system getting data from over the World Wide Web when congestion slows the Web's transmission of packets. While timeliness is widely recognized as important in business rules in real-time systems, timeliness is also usually valued in other types of systems.

One way to visualize the contribution of business rules in systems is to regard them as state-changers. That is, when a business rule is fired (executed) during a system's performance, a change in the state of something usually occurs. As the system runs, but before the rule is fired, the system is in some specific state. After the rule is fired, the system is either in the same state or has moved into a different state, depending on which action portion was selected by the decision portion of the rule. A statechart diagram such as used in unified modeling language (UML) can represent state changes; see Figure 7 for a simple common example illustrating interactions among ten business rules (Rumbaugh, Jacobson, & Booch, 1999). This can help focus attention on what the business rule does in and for the system, expressed in specific operational terms. It can also depict what the system does in and for the organization.

Visualizing changes in the business rules, the system, and the organization can be an aid both for an organization's management and for the systems personnel. To implement each change in an organization requires changes to be made in at least one system and usually changes in several systems. To implement each change in a system requires changes to be made in at least one business rule and usually in several business rules. At each level, boundary changes may be made.

These change processes also work in the other direction, but with more opportunity for shortfalls, oversights, and impaired organizational performance. Thus, each change made in a business rule makes changes in at least one system and may result in consequential changes in other systems. Each change made in a system makes changes in at least one organization and may result in consequential changes in other organizations. At each level, boundary changes may be made. Introducing a change in a business rule is often regarded in practice as all that is needed to change an organization's operational performance. While only making such a change usually does make a difference in an organization's operational performance, it often has unintended side effects because of ripple changes warranted but not made in the combination of services in the system or in the boundaries. These unintended side effects often result in some impairment in an organization's operational performance—i.e., its health and fitness.

The most common unintended side effects are on the service of personnel. Any change in business rules has the potential to change what people do or how and when they do it in the system, and with whom they interact. When the potential becomes an actual change affecting people, then some of those people are losers and some are winners in the organization. Much of the internal political chaos and

Figure 7: An example of a UML statechart diagram for a left-turn (LT) signal light subsystem that is part of the traffic light control system at a four-way intersection

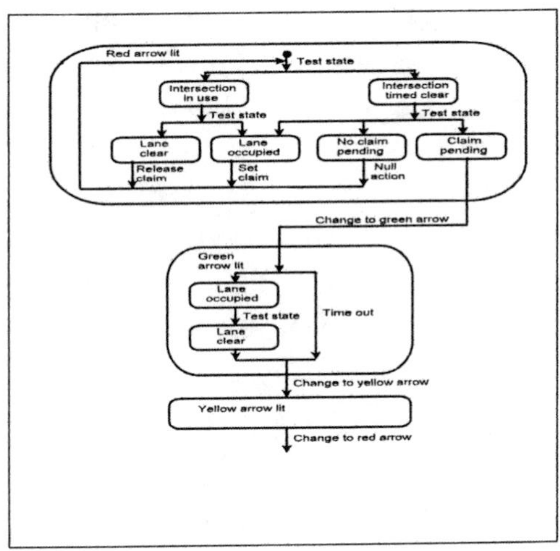

advocacy for change or for status quo in organizations springs from people seeking to be winners or seeking to avoid being losers. To make anyone in an organization be a winner or a loser, their personal performance in applying business rules in systems in the organization has to change relative to other persons', or the business rules or systems have to change.

The winners usually accept the change as appropriate and worth doing. Almost all people in organizations have experienced that to gain in position, prestige, recognition, promotion, importance, desirable transfer, upgrade, more favorable associates, increased remuneration, work space, access to resources, power, or other "perks," specific changes have to be made to the business rules or systems (i.e., the way the organization works) in order to give them a more significant role. For example, consider the case of a man working as a receptionist and assigned to do routine data entry of inventory corrections as the background fill work. A systems change is proposed that would have him keep the appointments books for a group of executives and move part of the data entry work to a clerical pool. The man welcomes this proposal, because he wants to get a better basis for bidding later on moving up to being a "go for" executive assistant to some one of the executives in the group, in order to give him a less sedentary job with more variety.

The losers normally resist the change and fight it both passively and actively, directly and indirectly. The loses can take many forms, many of which are important to some people, but are not important to others. For example, a particular change extends an organization's use of an enterprise resource planning (ERP) system, eliminating on a week's notice the job of one of the assistant purchasing agents. The affected man is offered, with no change in salary, a transfer to an incoming quality inspection group or an opportunity to train to become a sales associate. The man regards both choices as "a slap in the face." He regards the training offer as being forced out of a group he knows and feels comfortable with, into associating with a bunch of people he believes he has little in common with, all without any assurance of passing the training. He regards the transfer offer as a move from indoor office work to work in a warehouse, where he is taking orders and working with things instead of what he likes to do—trying to get people to do things his way. In an attempt to avoid the choices, he brings up with his present supervisor how the purchasing department will need help in adapting to the ERP system and adapting the ERP system to the needs of the department—and requests he be assigned to that work (and thus initiates a proposal for additional consequential change).

SOFTWARE MAINTENANCE AND EVOLUTION

As is common knowledge, information systems are currently very widely encountered in all but very small organizations. Historically, most information systems operated concurrently and were implemented primarily with personnel services using only manually operated machines, such as the abacus, filing cabinet, pen and ink, calculator, coin sorter, etc. With the advent of computers to serve in the machine role, some concurrency was lost because most computers multitask rather than run in a true concurrent manner. At present, an organization usually has to run information systems on different computers to get full concurrency in performance for the organization. In either case, however, information systems have not lost the personnel service contribution with its concurrency. What has happened is that the character of the personnel service contribution has shifted away from the routine processing and conversion of data to relatively more decision making and more using of the output data produced by "computerized" information systems as a data source. Such computerized systems use software for much of their implementation of the systems' methods. To make changes in those information systems hence often requires changing the software along with the other characteristics of the systems—a process usually termed ***software maintenance***.

For some decades, software maintenance has been regarded from the point of view of the intentions of the personnel affected by the system (Swanson, 1976). An example is "perfective" maintenance. A modern trend is to focus on the activities and processes involved in software maintenance projects and to characterize the maintenance in terms reflecting what is actually done (Kitchenham et al., 1999; Chapin, Hale, Khan, Ramil, & Tan, 2001). An example is "enhancive" maintenance. Managing maintenance in terms of intentions is much less objective and much more intangible and insubstantial than managing maintenance in terms of what people have done and are to do. People's intentions are more difficult for management to identify, measure, and assess than are people's performances on the job, and it is performance—not intentions—that change the health and fitness of organizations. For systems and organizations also, intentions are trivial, and performance is critically important in changing organizational health and fitness.

Over the decades that we have had computers in systems and especially in information systems, we have learned how to design and implement the systems to make the software of those systems more easily, rapidly, and accurately understood by the systems personnel (an example is Prowell, Trammel, Linger, & Poore, 1999). We have also learned activities and processes to make the software quite reliable and modifiable relatively quickly and at a lower cost.[4] Precisely specified boundaries in terms of data, hierarchical or table-driven functionality, restricted data access, locally recognized and accepted patterns or plans or templates, specific attention to the timing of data value changes, and consistent current accurate documentation are high on the list of attributes contributing to good software maintainability. In doing software maintenance, we normally strive to leave the software cleaner on all such attributes than it was before the maintenance was done.

All that has been noted previously in this chapter about changes to organizations, systems, and business rules applies also to software maintenance. This is because software is used in implementing systems that use computers for part of their machine services. One trend in the software used in systems that interact with people, such as point-of-sale systems and accounting systems, is facilitating the human interface. The proportion of the software devoted to that interface has been increasing and now often exceeds half of the software in a system. Changing the software to facilitate or improve the friendliness of the human interface is often treated as a means of modifying the services of personnel in the system. However, strengthening or facilitating the human interface may actually be like "frosting on the cake" with respect to the needed changes in personnel services in the system.

The processes of software maintenance are not limited to modifying just the software:

- Modifying the software typically modifies part of the machine services in a system. This in turn typically requires rebalancing the contributions from the non-computer service providers in the system, if the organization's operational performance is to be preserved or improved. Failure to make the rebalancing or readjusting changes appropriately typically results in some impairment of the organization's operational performance. Software maintenance is more than tinkering with the code; rather, it is managing change in the organization (Canning, 1972). If management would preserve or improve the organization's operational performance, then management has a stake in how the software maintenance is directed and what is achieved.
- Not all software maintenance involves making changes only to the software implementations of business rules. Some software maintenance work, especially that initiated by the systems personnel, results in modifying software properties usually undiscernable to management. Examples are changes in the speed of system execution, the amount of computer storage used, the involvement of operating system services, the arrangement of data in a database, the transition to a new version of a directory, etc. Such changes affect primarily the methods, machines and materials services in the system, but rarely result in changes to the personnel services.
- Some software maintenance changes the system's documentation. Except for user manual changes, such changes usually directly affect relatively few people in an organization other than the systems personnel. When the documentation is changed to be more accurate, current and relevant, the work of the people directly affected is nearly always improved, and indirectly—through faster, better and lower cost changes to systems, the organization's operational performance is subsequently improved.

- The largest part of software maintenance does not modify the software—i.e., it leaves the code and the documentation unchanged. What it does is use the code and the documentation to:
  1. build or refresh the systems personnel's knowledge and understanding of the system and its software;
  2. support consultation with users and other affected personnel about the use and characteristics of the system and its software, such as about its contribution in the organization, how this system fits with other systems, how various classes of transactions are handled by the system, etc.;
  3. validate or test the software as a whole to confirm its operational performance (regression testing is an example);
  4. provide a basis for answering questions from management about the system and its software, such as "What would be the likely time and cost to make a particular change in this system?";
  5. act as a vehicle or aid in the training or refreshing of people in the use of the system and its software, such as by teaching, on-site tutoring, staffing a "help desk," etc.; and
  6. serve as a basis in helping management devise ways to give the organization new or changed functionality (function shifting) through different and changed uses of the existing systems.

  More systems personnel time goes into these six activities than into any other of the activities and processes in doing software maintenance. Rebalancing services, making changes in systems' properties and functionalities, upgrading the documentation, and the six activities just noted above, can all contribute to preserving and improving an organization's operational performance capability.

Software maintenance is also the operative vehicle for accomplishing software evolution (Chapin et al., 2001). The activities and processes in software maintenance that implement changes in the business rules are the primary means of accomplishing software evolution. This is because software evolution results in new versions of computerized information systems provided for use by the personnel in the organization.

Management often directs, on an ad hoc basis, that software maintenance be done to implement changes in business rules. The usual result of such ad hoc direction is unplanned, uncoordinated, and episodic changes and growth in the software and systems used by the organization. Taken together, such "random walk" changes result in unmanaged software evolution and hence in unmanaged system evolution. Haphazard improvement or impairment in the organization's operational performance may result. Alternatively, management can forecast and specify a deliberate planned and targeted set of steps to guide the software evolution to strengthen the organization's operational performance—i.e., the organization's health and fitness. Such planned evolution has to be periodically redirected as the organization's environment and milieu change in unpredicted extents or ways. In carrying out such targeted software evolution, software maintenance (especially on properties and documentation) may be done in advance of management-specified need, on a preplanned or scheduled basis.

THE POTENTIAL OF SOFTWARE MAINTENANCE

Negatives

The major negative potentials relating organizational health and fitness with software maintenance are not new—they are some of the lessons learned over the years by organizations doing software maintenance. They have been exerting their influence for decades, and account for most of the under recognition of the contribution of software maintenance in organizations. Ten activities with negative potential are:

1. Management in the organization does not regard improving the fit of the current systems to the current environment and milieu as a rewardable use of personnel time, effort or skills. Software maintenance done, regardless of how effectively, typically results in no promotions, no salary increases, no non-niggardly bonuses, etc. Getting such rewards in most organizations is far easier and more likely by proposing discarding and replacing one or more existing systems with something new and untried in this organization. This management predilection persists in spite of evidence that discarding and replacing usually is not the road to improved organizational performance (Jones, 1993).
2. Management does not usually and routinely quantify the degree of misfit between the operational performance of the current systems and management's view of what their operational performance should be. Some degree of misfit is almost inevitable, but a major misfit can result in a major degradation in the organization's operational performance (Drucker, 1973b). Such misfits are most likely to arise from deficiencies in the internal stimuli-responses network in the organization, usually because the stimuli (often as metrics) are not available or are ignored, or because the linking network only works partially, or because management elects to respond to available stimuli with action other than directing and supporting software maintenance.
3. Management keeps software maintenance out of the official organizational chart, assigns responsibility for software maintenance to no one, red-lines out of budgets any activities labeled as software maintenance, and advocates

among associates and other managers that the organization should pare the money put into that "rat hole" of existing systems (Chapin, 1993). A starved, neglected, ignored, and invisible or frowned upon activity can help motivate an organization's personnel to look for other ways to preserve or improve an organization's operational performance.

4. Management accepts the view that the proper province of software maintenance is only diddling with the code. Management rejects the view that software maintenance is the management of change in the organization (Canning, 1972). The usual consequence is that the organization's operational performance is impaired, and management usually blames either the current systems as being incompetent to some degree, or the software maintenance as not being properly done, or both.
5. Management manages software maintenance reactively (Chapin, 1985). Management regards software maintenance requests or needs as unwelcome, unsought situations that are best minimized and handled by firefighting and damage control. Something broken or worn out might be something that management might consider having maintenance done on to return it to its prior level of functionality. Unlike machines, software does not wear out and very rarely gets broken, and hence, in management's view, should very rarely receive any maintenance attention. And if it should become obsolete, then the appropriate management action is to replace it and gain the benefits of improved technology.
6. Management sets and enforces deadlines for the completion of software maintenance work based upon the requesters' expressed preferences. Management typically treats the systems personnel's estimates of the scheduling, time, effort, and resources needed to do the requested work as self-serving padded data unworthy of serious consideration. To reduce any gap between the deadlines proposed by requesters and by the systems personnel, management applies pressure to the system personnel to meet the requesters' deadlines (Couger & Colter, 1985).
7. Management declines to recognize that software maintenance is a more demanding job than software development, in spite of supporting evidence (Chapin, 1987). Hence, management often assigns junior-level and new-hire personnel to software maintenance and sets the pay scale to pay less to personnel normally assigned to software maintenance than to software development. The result is an effectively higher unit cost for most quality levels of software maintenance done, that in turn discourages management from having software maintenance done.
8. Management views software maintenance narrowly, such as making corrections or enhancement, or both, to the source code. Management regards such

activities as studying the documentation, training user personnel, regression testing, replacing existing algorithms with more elegant or rapid ones, upgrading documentation, etc., as unnecessary "frill" overhead to be minimized when done by systems personnel (Lietz & Swanson, 1980). The result is an increasing cost per unit of doing software maintenance (even as narrowly regarded) and impaired operational performance for the organization as software maintenance lags more and more.

9. Management elects ad hoc software evolution as the norm in the organization, rather than planned and scheduled software evolution. Management ignores or downplays having software follow a "career path" even when it may embrace career pathing for employees. The result is usually having to play and pay for catch up or to attempt to do system replacement when the software has been allowed to stagnate (Lehman & Belady, 1985). Stagnated software usually indicates that the system falls short because of accumulated misfits in providing the desired operational performance, with the organization's needs to fit with the current environment and milieu.
10. Management avoids adopting or using the activities with positive potential or adopts defenses against them to keep them in check and offset or mitigate the effectiveness of software maintenance in contributing to preserving or improving the organization's operational effectiveness.

Positives

The major activities with positive potential relating organization health and fitness with software maintenance are also not new—they too are some of the lessons learned over the years by organizations doing software maintenance. While often overwhelmed by the negative potentials, in practice, the positive potential activities have contributed to achieving some effective software maintenance and improving the operational performance of some organizations, or offsetting or defeating some of the negatives to some degree. The positives get far less management attention than the negatives, primarily from the typical short planning horizon most managers currently support ("We have to look good this quarter.") and from personnel pressure to permit and reward solo stars rather than competent stable teams of personnel. Ten activities with positive potential are:

1. Management acts to eliminate, neutralize, or reverse the activities with negative potential to clear the way for the activities with positive potential to be effective in contributing to preserving or improving the organization's operational effectiveness.
2. Management takes toward a proactive software evolution approach software management. This involves the relationships between the management

personnel and the systems personnel and the personnel using the system. Communication and consultation between users, management and the systems personnel about anticipating changes and defining needs and their timing are usually helpful. Planning out the staging and transitions for a series of changes helps reduce the personnel's feeling of chaos and "fud" (fear, uncertainty, and doubt) and build feelings of assurance and confidence among the personnel affected by the changes.

3. Management publicizes the stable low-cost, personnel-empowering way that change via software maintenance is handled in the organization, as well as the actual track record from past and current software maintenance projects. Management also contrasts this with the typical chaotic, high-cost fud characteristics and the poor industry track record of discarding existing systems and replacing them with new systems. In systems change, as the software is maintained (or replaced), the winners rarely need their gains glorified, but the losses for the losers have to be minimized, as by skill-upgrading training, attractive assignments, etc., to avoid the performance impairments from personnel turnover and reduced morale or motivation.
4. Management encourages maintenance teams and the use of repeatable processes. The teams may be small, as in pair programming, to ten members or more (Biggs, 2000; Humphrey, 2000; Prowell et al., 1999). But the processes the teams use are known and understood by all of the team members, even when some of the team members may not be expert in their use. The processes are also repeatable, so that nearly identical results are achieved regardless of which team does the work. The teams apply quality assurance techniques and methods built into the ways they work, rather than leaving quality assurance as a separate "may be done well" step near the end of the maintenance project. An independent, separate software quality assurance (SQA) activity may also contribute value to some software maintenance projects.
5. Management encourages the fractal-like characteristic of business rules, systems, and organizations to help evaluate the quality of the results of software maintenance processes (Mills, Linger, & Hevner, 1986). Data about this fractal-like characteristic has been scattered throughout this chapter (for example, compare Figures 1, 3, and 5). Deviation from the fractal-like characteristic appears to be associated with inelegance and inefficiency in the level at which it appears and in any levels above. The impairment that arises stems from the deviation's effect upon the software maintenance processes, and tends to be cumulative. However, software maintenance can be done in ways that encourage and utilize the fractal-like characteristic.

6. Management at the first-line supervisory level makes assignments to the team members and receives assignment completion data in the form of documents. These documents may be in textual, graphic, numeric, or code styles, and in form may be on paper or in an electronic and often online form. When in an online form, the documents may include voice recordings. Whatever form the documents take, working from documents to produce documents provides management with evidence of what was done by whom and when, and at what quality level (Chapin, 1965–1990).
7. Management insists upon the correct and coherent use of the relevant software engineering technologies in software maintenance. Software engineering is in general not widely practiced well, even in software development, (as the large proportion of Level 1 Capability Maturity Model [CMM] software development sites testifies), and even less well practiced in software maintenance. Almost any tool can be misused and, when misused, the misuse usually rots rather than builds the target result. Three frequently hyped software engineering technologies that are very often misused in software maintenance are:
   - the object-oriented technologies—only a very careful use of specific forms avoids reducing software maintainability and increasing the unit cost of doing software maintenance (Li & Henry, 1995);
   - the structured technologies—very often claimed to be widely and correctly used, but actually observed to be commonly misused or inconsistently used, thus making software harder to understand for the systems personnel (Chapin, 1978); and
   - the software reuse technologies—these promising technologies have so far not lived up to their potential in software maintenance, primarily because of dictionary and repository factors (McClure, 2001).
8. Management enforces consistency and completeness in the definition and enforcement of boundaries for business rules, systems, and organizations. Overlapping, conflicting, partial, indefinite, and transitory boundaries, or the presence of unbounded items (items not included within any appropriate level boundary) contributes to confusion and inconsistent action and thus impairs operational performance. Boundary definition, even when aiding software tools that have been available, has gotten little explicit attention except when a question has been raised about a specific item; e.g., "On what boundary if any is this item?" Documenting boundaries is aided by the use of comprehensive data dictionaries and some other computer-assisted software engineering (CASE) tools or, in their absence, explicit annotation in the documentation.
9. Management directs that the software be kept well groomed. Software is well groomed when the documentation matches the code and both conform fully

to the local standard operating procedure (SOP). This usually includes data consistently named and described, with clear and meaningful annotation placed appropriately in the code, and with accurate and frequent cross-reference between the code and the non-code documentation. Full test plans with their test description, test directions, test input data or how to find them, and correct test output data for test run comparison or how to find them are often included as part of well-groomed software. Well-groomed software is easier for the systems personnel to understand and quickly find in it what they want to find, thus reducing unit cost for software maintenance, speeding maintenance, and improving the quality of the software maintenance.

10. With each request for software maintenance, management requires from the requester a quantitative estimate (in monetary and other quantitative terms) of the value to the organization of the requested software maintenance. Excluded from the estimate is the estimate of the cost to the organization for the actual doing of the requested software maintenance, since that estimate is made separately by the systems personnel and is used to get the net value of the requested maintenance. Then the organization management and the systems personnel compare the value estimate with the cost estimate to get a benefit/cost estimate, including the time value of money, to assist the management in deciding whether or not to do the requested software maintenance. If the maintenance is done, then about a year subsequent to the initial use of the modified software, an audit is made to determine the actual realized benefit/cost. The realized benefit/cost data assist in making future estimates of value and cost, and establish the real quantitative value to the organization of doing software maintenance.

Maintenance Maturity

The twenty activities just listed, half positive and half negative, really are just some highlights or aspects of what is often called "process maturity." The general concept of process maturity as applied to information systems has been most widely publicized in the work of the Software Engineering Institute (SEI) with its Capability Maturity Model (CMM) for software development (Humphrey, 1995). The SEI CMM gives scant attention to software maintenance in any of its five levels of maturity. This oversight is being rectified by others; for example, for corrective maintenance, see the CM3 described elsewhere in this volume by Dr. Mira Kajko-Mattsson.

In summary and in generic form, an organization operating at a low level of maturity in software maintenance usually displays most of the following characteristics:

- the program and system documentation is weak, incomplete, not up-to-date, inaccurate, or hard to understand, and user manuals usually share the same states;

- the personnel find they rely mostly on the source code and their own memories to help them understand the software;
- the personnel usually work solo but are often specialized to certain software or certain processes (such as preparing test runs), commonly with little variety;
- work is unplanned and often done informally in response to verbal requests, and a separate quality assurance activity is rarely used;
- deadlines drive the work, and methods and techniques are used in an ad hoc manner with little repeatability;
- software configuration management and software quality assurance are absent or inconsistently applied;
- the personnel career paths are ambiguous, and supervisors rarely spend time with subordinates except to ask about the meeting deadline status;
- personnel performance is assessed in qualitative informal terms, and contact with the software's stakeholders is fleeting and rare; and
- in general, management has only a weak ability to affect outcomes.

In summary and in generic form, an organization operating at a high level of maturity in software maintenance usually displays most of the following characteristics:

- each of the personnel have expert skills in some areas, but are able to work in nearly any team because accurate, clear, readable documentation provides data about the systems to be worked on;
- the personnel doing the maintenance consider themselves as professionals applying their expertise as appropriate to whatever may be the situation;
- for most types of situations faced, the personnel know and use a gradually evolving software maintenance SOP that includes some form of Software Configuration Management (SCM) and either some form of regression testing, or Independent Verification and Validation (IV&V) or inspection or audit, or some combination as called for in the SOP;
- some software reuse is practiced with the aid of an active data dictionary or software repository or both;
- updated documentation is a byproduct of the performance of all work that changes any characteristics of the requirements, source code, object code, components, or interfaces, or any of their equivalents;
- the software maintenance work requests and backlog are managed to support orderly software evolution; and
- career paths for the personnel may be either technical or managerial, and supervisors are primarily concerned with quality of service, planning, and resource availability and deployment.

CONCLUSION

In this chapter, "organization" has been the term for a human-created, human-operated entity for a human-designated purpose. Organizations usually have suborganizations as components. Although such parallels are perilous, comparing eight activity attributes of living things and of organizations makes us aware of some key characteristics of organizations, and of what contributes to health and fitness as manifested in the operational performance of the entity or organization.

In organizations, systems determine how recurrent situations get handled by the methods that combine the services of personnel, machines and materials in order to get them to operate together. Information systems are of special interest in this chapter since they often involve software that implement the methods. In a system, a key part of the methods is management's choice of the business rules. These have two main portions, decision and action. Business rules typically are nested and implemented within systems, and systems typically have subsystems as components. The systems and their business rules are the primary contributors to the operational performance capability of an organization—i.e., to the organization's health and fitness.

To change an organization's health and fitness, management has to specify changes to the existing repertoire of systems and how those system operate. Typically, changing an existing system requires changing the business rules within the system, often with supporting or consequential changes in the services provided by personnel, machines, materials, and methods. The personnel changes are typically the most sensitive and complicated to handle, and can result from making changes to the business rules, to a system, and to an organization, due to the fractal-like characteristics of systems and their components.

Management has a key role in selecting, directing, and specifying change in an organization, a system, or a business rule. Implementing management-required changes often requires the systems personnel to make additional ripple, supporting, or consequential changes. In information systems, the systems personnel normally take care of these additional changes that may sometimes be extensive. Computerized information systems get changed mostly through software maintenance processes, yet information systems involve more than software—they also involve personnel, machines (in addition to computers), materials, and methods. This emphasizes that the management of software maintenance involves more than just managing the changing of software—software maintenance makes changes to a system's and hence also to an organization's operational performance. In closing, this chapter highlighted twenty lessons learned. Ten are software maintenance activities that can have a negative potential, and ten are software maintenance activities that can have a positive potential on the health and fitness of an organization.

ENDNOTES

1 Except where otherwise indicated, this chapter draws primarily upon the author's observations and experiences, and upon these sources: Chapin, 1955; Chapin, 1971; and Chapin, 1965–1990.

2 The intended distinction between "environment" and "milieu" in this chapter is that the environment is the natural physical surrounding context of the entity, and the milieu is the human-dependent, cultural and social-institution surrounding context of the entity.

3 This is a minor variation on an engineering definition. A discussion of some of the varied meanings of the term "system" can be found in Klir (1991).

4 This involves many technical considerations. Two examples are Chapin (1978) and Chapin (1999).

REFERENCES

Biggs, M. (2000). Pair programming: Development times two. *InfoWorld, 22*(30) (July 24), 62, 64.

Canning, R.G. (1972). That maintenance 'iceberg.' *EDP Analyzer, 10*(10), 1–14.

Chapin, N. (1955 & 1963). *An Introduction to Automatic Computers*. Princeton, NJ: Van Nostrand Co.

Chapin, N. (1965–1990). *Training Manual Series*. Menlo Park, CA: InfoSci Inc.

Chapin, N. (1971). *Computers: A Systems Approach*. New York: Van Nostrand Reinhold Co.

Chapin, N. (1971). *Flowcharts*. Princeton, NJ: Auerbach Publishers.

Chapin, N. (1974). New format for flowcharts. *Software Practice and Experience, 4*(4), 341–357.

Chapin, N. (1978). Function parsing in structured design. *Structured Analysis and Design Volume 2,* (pp. 25–42) Maidenhead, UK: Infotech International, Ltd.

Chapin, N. (1985). Software maintenance: A different view. *AFIPS Proceedings of the National Computer Conference* (Vol. 54, pp. 507–513) Reston VA: AFIPS Press.

Chapin, N. (1987). The job of software maintenance. *Proceedings Conference on Software Maintenance–1987* (pp. 4–12) Los Alamitos, CA: IEEE Computer Society Press.

Chapin, N. (1993). Software maintenance characteristics and effective management. *Journal of Software Maintenance, 5*(2), 91–100.

Chapin, N. (1999). Coupling and strength, a la Harlan D. Mills. *Science and Engineering in Software Development: A Recognition of Harlan D. Mills' Legacy* (pp. 4–13) Los Alamitos, CA: IEEE Computer Society Press.

Chapin, N., Hale, J.E., Khan, K.M., Ramil, J.F., & Tan, W.-G. (2001). Types of software evolution and software maintenance. *Journal of Software Maintenance and Evolution, 13*(1), 3–30.

Couger, J.D., & Colter, M.A. (1985). *Maintenance Programming*. Englewood Cliffs, NJ: Prentice-Hall.

Drucker, P.F. (1973a). *Management: Tasks, Responsibility, Practice* (pp. 95–102, 517–602) New York: Harper &Row.

Drucker, P.F. (1973b). *Management: Tasks, Responsibility, Practice* (pp. 430–442) New York: Harper &Row.

Humphrey, W.S. (1995). *A Discipline for Software Engineering*. Reading, MA: Addison Wesley.

Humphrey, W.S. (2000). *Introduction to the Team Software Process*. Upper Saddle River, NJ: Prentice-Hall.

Jones, T.C. (1993). *Assessment and Control of Software Risks*. Englewood Cliffs, NJ: Prentice-Hall International.

Kitchenham, B.A., Travassos, G.H., Mayrhauser, A.v., Niessink, F., Schneidewind, N.F., Singer, J., Takada, S., Vehvilainen, R., & Yang, H. (1999). Towards an ontology of software maintenance. *Journal of Software Maintenance, 11*(6), 365–389.

Klir, G.J. (1991). *Facets of System Science* (pp. 3–17, 327–329) New York: Plenum Press.

Lehman, M.M., & Belady, L.A. (1985). *Program Evolution: The Process of Software Change*. New York: Academic Press.

Li, W., & Henry, S. (1995). An empirical study of maintenance activities in two object-oriented systems. *Journal of Software Maintenance, 7*(2), 131–147.

Lientz, B.P., & Swanson, E.B. (1980). *Software Maintenance Management*. Reading, MA: Addison Wesley.

McClure, C.L. (2001). *Software Reuse: A Standards Based Guide*. Los Alamitos, CA: IEEE Computer Society Press.

Mills, H.D., Linger, R.C., & Hevner, A.R. (1986). *Principles of Information Systems Analysis and Design*. Orlando, FL: Academic Press.

Prowell, S.J., Trammell, C.J., Linger, R.C., & Poore, J.H. (1999). *Cleanroom Software Engineering: Technology and Process*. Reading, MA: Addison Wesley Longman, Inc.

Rumbaugh, J., Jacobson, I., & Booch, G. (1999). *The Unified Modeling Language Reference Manual*. Reading, MA: Addison Wesley.

Swanson, E.B. (1976). The dimensions of maintenance. *Proceedings of the 2nd International Conference on Software Engineering* (pp. 221–226) Long Beach, CA: IEEE Computer Society Press.

Chapter II

Problem Management within Corrective Maintenance

Mira Kajko-Mattsson
Stockholm University & Royal Institute of Technology, Sweden

Problem management is the dominating process within corrective maintenance. It not only handles software problems but also provides quantitative feedback important for assessing product quality, crucial for continuous process analysis and improvement and essential for defect prevention.

Chapter 2 gives a detailed description of the problem management process and its constituent phases, activities, and roles involved. It also provides (1) general definitions on the basic terminology utilized within corrective maintenance, (2) roadmaps visualizing the corrective maintenance from different perspectives, (3) general descriptions of the most important issues within support (upfront maintenance) process – the process tightly collaborating with the problem management process, and finally, (4) it gives a glimpse into Corrective Maintenance Maturity Model (CM^3): Problem Management – a process model specialized to only problem management process within corrective maintenance.

INTRODUCTION

Corrective maintenance is a very important process for achieving process maturity. It not only handles the resolution of software problems, but also provides a basis for quantitative feedback important for assessing product quality and reliability, crucial for continuous process analysis and improvement, and essential for defect prevention and necessary for making different kinds of decisions. Yet, the domain of corrective maintenance has received very little attention in the curriculum within academia or industry. Current literature provide very coarse-grained descriptions. Extant process models dedicate at most one or two pages to describing it.

In large *software* organizations, a corrective maintenance process is seldom monolithic. Instead, it is a complex family of various processes collaborating with each other, where each process executes a clearly defined task. In this chapter, we present one of the main processes within corrective maintenance – problem management process. We also illuminate the most important issues within support (upfront maintenance) process – the process tightly collaborating with the problem management process.

The first section, "Software Maintenance," gives a general introduction to the definition of maintenance and it's categories. It also illuminates background and problems concerning these definitions. The next section, "Corrective Maintenance," presents corrective maintenance and its basic terminology. The sections, "Maintenance Organizations," "Product Perspective," and "Life Cycle Perspective" visualize the corrective maintenance process from the organizational, product, and life cycle perspective, respectively. Then, "Problem Management at the Maintenance Execution Process Level" describes problem management process, its phases, and roles involved in its execution. Finally, the last section, "Corrective Maintenance Maturity Model: Problem Management," gives a glimpse into Corrective Maintenance Maturity Model (CM^3) – a process model specialized to only corrective maintenance. Our description of problem management in this chapter is based on the theory extracted form CM^3 (Kajko-Mattsson, 2001a).

Software Maintenance

What is corrective maintenance and how does it relate to other types of maintenance work? To explain this, we must place corrective maintenance within total maintenance and identify its relation to other maintenance categories. The majority of the software community follows the IEEE suggestion for defining and categorizing maintenance. As depicted in Table 1, the IEEE defines software maintenance as "the process of modifying a software system or component after delivery to correct faults, improve performance or other attributes, or to adapt to a changed environment" (ANS//IEEE STD-610.125 1990).

Not everybody agrees on this definition today. There prevails a controversy on the choice of maintenance scope, its constituents, time span, and on drawing a dividing line between software development and software maintenance (Chapin, 2001; Kajko-Mattsson, 2001d; Kajko-Mattsson, 2001f; Parihk, 1986; Schneidewind, 1987; Swanson, 1999). This is clearly visible in so widely varying cost estimates of software maintenance, spanning between 40%-90% of the total software life cycle cost (Arthur, 1988; Boehm, 1973; Cashman & Holt, 1980; DeRoze & Nyman, 1978; Glass & Noiseux, 1981; Jones, 1994; Mills, 1976).

According to Martin and McClure (1983), Pigoski (1997), and Schneidewind (1987), the definition suggested by the IEEE states that maintenance is entirely a postdelivery activity. No consideration is being made to the predelivery phase of the software life cycle, despite the fact that successful postdelivery/postrelease maintenance phases depend highly on the degree of engagement of maintenance organizations during the predelivery life cycle phase. Hence, maintenance should start earlier than after delivery.

Irrespective of who is going to maintain the software system (either a development or a separate maintenance organization), certain maintenance activities must be conducted during predelivery (development) or prerelease (evolution/maintenance) phases. These activities are for the most part of a preparatory nature for the transition and postdelivery/postrelease life-cycle phases. They concern planning for postdelivery/postrelease maintenance, providing for its cost-effectiveness, designating and creating a maintenance organization, allowing maintainers to learn the product, and most importantly, allowing them to affect and control the process of building maintainability into a software system.

To add zest to the problem of defining maintenance, there is also a great deal of confusion concerning the choice and definition of maintenance categories. Protests have been raised against the IEEE definitions. As presented in Table 1, the IEEE has designated four maintenance categories. They are corrective, perfective, adaptive, and preventive. This categorization, however, is not mutually exclusive (Chapin, 2001). Our study has revealed that definitions of perfective and preventive maintenance categories overlap (Kajko-Mattsson, 2000; Kajko-Mattsson, 2001f). In addition, these definitions are too general, leading to subjectivity and confusion within the maintenance community (Chapin, 2001; Foster, Jolly, & Norris, 1989). According to Chapin (2001), the IEEE classification of maintenance categories is neither exhaustive nor exclusive. Maintenance requests cannot be reliably and consistently classified from available objective evidence. The same maintenance

Table 1. IEEE definitions of software maintenance (ANSI/IEEE STD-610.12, 1990).

IEEE 90, Std 610.12-1990

Maintenance:(1) The process of modifying a software system or component after delivery to correct faults, improve performance or other attributes, or adapt to a changed environment. (2) The process of retaining a hardware system or component in, or restoring it to, a state in which it can perform its required functions. *See also*: **preventive maintenance**.

Corrective maintenance: Maintenance performed to correct faults in hardware or software.

Adaptive maintenance: Software maintenance performed to make a computer program usable in a changed environment.

Perfective maintenance: Software maintenance performed to improve the performance, maintainability, or other attributes of a computer program.

Preventive maintenance: Maintenance performed for the purpose of preventing problems before they occur.

task can be differently classified (e.g., either as corrective or perfective) depending on the intentions of the beholder.

Let us have a look at the IEEE definitions of maintenance categories and map them onto the meanings attached to them by the software community today. *Perfective maintenance* is mainly understood as the process of adding new features to the extant software system. An example of new features could be a new sorting function in an industrial robot or a new algorithm making this robot execute faster. *Adaptive maintenance* is mainly understood as the process of adapting the software system to the changed environment. For instance, some hardware/software parts of the industrial robot might be upgraded, and hence, some modification of software might have to be made to accommodate to the new hardware/software. *Preventive maintenance* is understood as the process of preventing software problems before they occur. This usually implies discovering software problems and attending to them prior to customer complaints (Chapin, 2001; Kajko-Mattsson, 2001f). Finally, corrective maintenance is the process of attending to software problems reported by the users of the affected software systems. For instance, an arm in a robot turns to the right instead of to the left.

When scrutinizing the IEEE definitions of the maintenance categories, it becomes apparent that the IEEE definition of perfective maintenance does not explicitly express its objectives. Enhancements, for example, are not stated explicitly as a constituent, but can be implicitly understood by inference from "improved performance" (ANSI/IEEE STD-610.12, 90; ANSI/IEEE STD 1219, 1998) and "other attributes" (ANSI/IEEE STD-610.12, 90) (see Table 1). There is also an overlap between perfective and preventive maintenance. As we can see in Table 1, the IEEE states that the improvement of maintainability is a constituent of perfective maintenance. However, the majority of the software community treats it as a constituent of preventive maintenance (Kajko-Mattsson, 2000; Kajko-Mattsson, 2001f).

Our conclusion is that the domain of maintenance is not unanimously and objectively understood today. The extant process models do not help in increasing objectivity. On the contrary, by proposing one generic process model for all maintenance categories, they are too general. Hence, they do not help in acquiring a deep understanding of the scope of each maintenance category. They do not offer the visibility required to each type of maintenance work. At most, they dedicate one or two pages to describing their differences. Covering such a huge domain by one and the same process model has led to many inconsistencies and unanswered questions.

How Can We Remedy the Problem?

One way to remedy this is to provide detailed and exhaustive descriptions of maintenance and its categories. Another way is to construct process models specialized to each maintenance category. Maintenance categories differ too much to be lumped together under one and the same model. We believe that each maintenance category deserves its own process model (a specialized model). This would offer a good opportunity to scrutinize the definition of each maintenance category, its name, and goal. Most importantly, this would lead us towards more objective understanding and measurement. Specialized models would also aid in obtaining agreement on a single classification system for each maintenance category and its inherent activities. They should be fine-grained enough to allow maximal visibility into the process. To the knowledge of the author of this chapter, there are no detailed process models explicitly defined for each maintenance category.

As a start, we have built a process model for corrective maintenance category. We call it *Corrective Maintenance Maturity Model (CM³)*. In this chapter, we are going to present problem management process – one of the most important processes necessary for managing software problems within corrective maintenance. It is based on one of the CM³'s constituent process models called *CM³: Problem Management* (Kajko-Mattsson, 2001a). Our goal is to provide a thorough description of problem management within corrective maintenance in order to provide guidance to industrial organizations in the process of building or improving their corrective maintenance processes, and to provide students and researchers with the sense of reality, size, and perspective to help them understand the spectrum of problem management within corrective software maintenance.

Corrective Maintenance

Corrective maintenance is the process of attending to software problems as reported by the users of the affected software systems. During corrective maintenance, we attend to all kinds of software problems: requirement problems, design problems, software code problems, user documentation problems, test case specification problems, and so forth. The main process within corrective maintenance is a problem management process encompassing the totality of activities required for reporting, analyzing and resolving software problems (Kajko-Mattsson, 2001a).

A software problem may be encountered either externally by an end-user customer or internally by anybody within a maintenance organization. A *Problem Submitter*, usually the person or organization who has encountered a problem, is obliged to report the problem to the maintenance organization. The maintenance organization must then attend to the reported problem and deliver the corrected version of the software system to the affected customer(s).

Due to the high cost of corrective maintenance, it is not an economical use of resources to attend to only one problem at a time. Instead, as depicted in Figure 1, maintenance organizations lump together several problems and attend to them in group. Before presenting the problem management process, we describe the basic terms relevant within the context of corrective maintenance: defect, fault, failure, software problem, and problem cause. For better visualization, we even model them in Figure 2.

Defect, Fault and Failure

A defect is an anomaly found in a software product. It can either be found in early life cycle phases during development, or later life cycle phases during maintenance. Examples of a defect might be a wrong algorithm, a spelling mistake, or misleading instructions in user manuals.

A subset of defects are faults. A fault is a defect found in executable software code. It corresponds to an incorrect specification of a program step, process, or data in a computer program (ANSI/IEEE STD-982.2, 1988; ANSI/IEEE STD-610.12, 1990; Florac, 1992). Another term for a fault is a bug.

A fault may never be encountered. But, if it is, then it leads to a failure. A failure is the departure of software operation from requirements or the non-performance of an action due to the fault. In our case, this would mean that our above-mentioned algorithm computes a wrong value.

Figure 1. Corrective maintenance process in general.

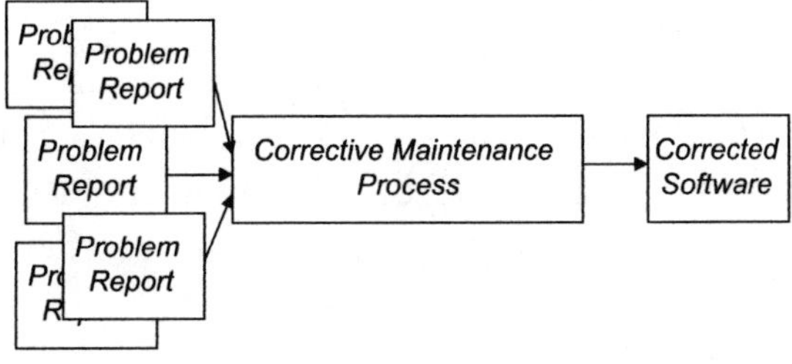

Figure 2. Basic concepts within corrective maintenance.[1]

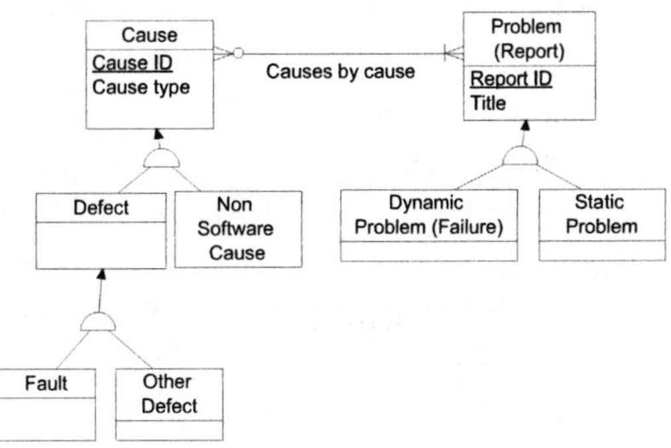

Software Problem

A software problem is a human encounter or experience with software that causes a doubt, difficulty, uncertainty in the use or examination of the software, or an encounter with a software failure (Florac, 1992). Other terms for a software problem are an incident or a trouble. Examples of a problem are the experiencing of a system crash, an encounter with a wrong value computed by a function, or an inability to understand documentation. It is software problems, not defects, that we report to the maintenance organization. At the problem encounter, we do not always know their underlying defects.

A software problem may be encountered dynamically during the execution of a software system. If so, then it is caused by a failure. A problem may also be encountered in a static environment when inspecting code or reading system documentation. In this case, a problem is the inability to understand code or documentation. An underlying defect to this problem might be misleading or unclear instructions.

Problem Cause

For the most part, problems are the consequences of defects. Sometimes however, problems may be due to something else, such as a misunderstanding, the misuse of a software system, or a number of other factors that are not related to the software product being used or examined (Florac, 1992). Additionally, in embedded systems, a problem could arise due to anomalies found in the hardware. Most of the software organizations today need to record causes of problems irrespective of their origin. For this reason, we define a cause as an origin of an imperfection or a flaw in a product (hardware or software defect) or a flaw in the operation of a product. An example of a problem cause might be a worn-out engine, choice of an inappropriate sequence of commands, or a software defect.

Maintenance Organizations

What does a corrective maintenance organization look like? Actually, this varies a lot. The scenarios may be the following:

1. Corrective maintenance is performed by one or several departments within a software organization. In this case, the maintenance department(s) perform different types of maintenance on the software product whose major parts have been developed by the same software organization.[2]
2. Corrective maintenance is conducted by several collaborating organizations, usually belonging to one and the same organizational group. These organizations both develop and maintain major parts of their software.
3. Corrective maintenance is outsourced to one or several organizations. These organizations do nothing but maintenance.

Irrespective of the number of the organizations involved in maintenance and their size, the total maintenance process is divided into two main logical process groups, upfront maintenance (support) and maintenance execution processes (see Figure 3). The upfront maintenance process mainly communicates on software problems with end-user customers, whereas the maintenance execution process designs and implements solutions to these problems, to then be delivered to the customers by the upfront maintenance (Kajko-Mattsson, 2001b).

Upfront Maintenance Process Level

At the upfront maintenance process level, usually called support, maintainers communicate on software problems with end-user customers. As depicted in Figure 3, the upfront processes are the Help Desk Process (HDP), and the Product Support Process (PSP). Their combined role is to maximize customer service by attending to all customer demands. Examples of these demands might be answering different types of questions on how to operate a software system, giving advice, receiving problem reports from customers and diagnosing software problems, suggesting solutions to these problems, if possible, or transferring them to the maintenance execution process level for further attention. Another role of upfront

Figure 3. Corrective maintenance—organizational perspective (Kajko-Mattsson, 2001b).

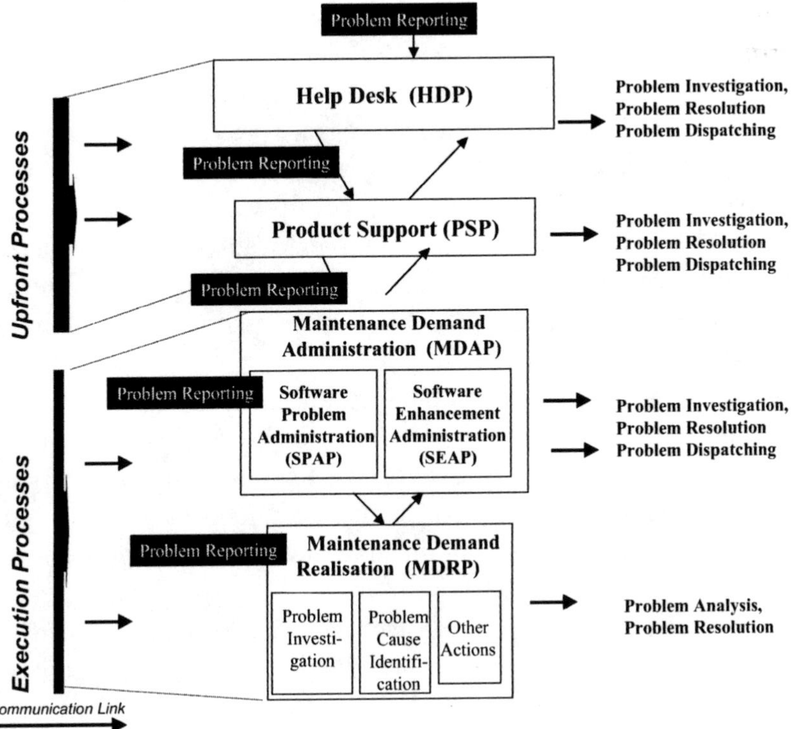

corrective maintenance is to relieve the already overstrained maintenance execution process (see Figure 3) from communicating with the customers thus helping them concentrate on the actual maintenance (modification) work (Kajko-Mattsson, Forssander, & Andersson, 2000a).

Most of the customer demands are attended to directly on the spot by the HDP. The HDP process is usually conducted by sales and support organizations. These organizations give support on the customer's site. Hence, they are closely located to the end-user customer organizations. Examples of the tasks at this level might be exchanging some hardware component, or an integrated hardware and software component, delivering an upgraded version of software, answering different types of questions, giving advice on how to operate the system, and identifying software problems.

At this level, the maintainers may not always be able (may not be competent enough) to attend to all the customer demands. In cases when they cannot help their customers, they turn to the maintainers at the PSP process level. The role of the PSP process is to provide technical expertise to the organizations at the HDP level, identify software problems, transfer them on to the maintenance execution process level, and supervise their management at this level.

Maintenance Execution Process Level

At the maintenance execution process level, maintainers attend to the reported software problems and modify software, if necessary. As depicted in Figure 3, there are two maintenance execution processes. These are the Maintenance Demand Administration Process (MDAP) and the Maintenance Demand Realization Process (MDRP). The MDAP analyzes maintenance demands and decides whether or not they should be attended to. Demands concerning software problems (corrective maintenance) are managed by the Software Problem Administration Process (SPAP), whereas the enhancements and adaptations are managed by the Software Enhancement Administration Process (SEAP). Both software problems and enhancements defined in the MDAP process may result in solutions being realized within the MDRP process.

During the SPAP process, maintainers administrate the reported software problem and design and plan for different tasks to be conducted at the MDRP process level. Examples of such tasks are investigating problems, verifying them, identifying problem causes, designing problem solutions, and implementing them. All these tasks are continuously administrated and monitored by the SPAP level.

Cooperation Between Upfront and Maintenance Execution Process Level

Due to the fact that many products have thousands of users geographically dispersed all over the world, there may be many upfront software maintenance organizations involved in supporting them. As depicted in Figure 4, one product that has been developed by one organization in Sweden may be attended to by up to 100 upfront maintenance organizations, or even more, as it is the case of ABB Automation Products (Kajko-Mattsson, 2001b). To provide good support to all customers, it is important to define a process that helps these organizations to co-operate. We should clearly specify the role and commitments of each maintenance organizational level. Also, we should create a communication pattern between customers and maintainers to make the maintenance process more effective and productive (Kajko-Mattsson, 2001b).

Communicating Problems from Customer to Maintenance Organization

To maximize maintainer productivity and customer satisfaction, maintenance demands should be communicated according to a particular pattern where each level constitutes a filter to the next maintenance level. In many organizations today, all customers first contact a maintenance organization at the HDP level (see Figure 3) (Kajko-Mattsson, 2001b). The customers should neither contact the PSP level nor maintenance execution level. This would infuse too much disturbance into the maintenance work and create chaos. Maintainers at the HDP level should manage all customer requests. The PSP maintainers are contacted only in cases when the demand concerns a software problem or it requires higher technical expertise. The main role of the maintainers at the HDP level is to filter out all trivial demands requiring less technical expertise. Their other main role is to filter out all duplicate problems, that is, the problems that have already been reported to the maintenance organization by some other customer(s). Only unique (previously unknown) software problems should be transferred to the PSP level.

Figure 4. Upfront maintenance organizations supporting one development and maintenance execution organisation in Sweden.

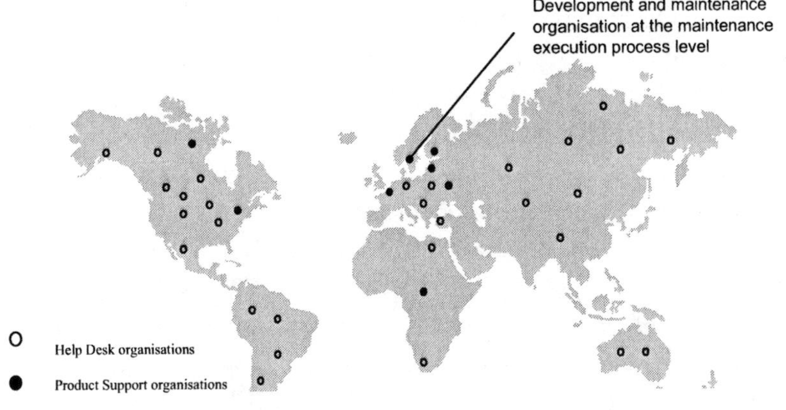

The PSP organizations should mainly deal with managing software problems. They should confirm that a software problem exists, check whether the problem is a duplicate, and if not, transfer it on for attendance to the maintenance execution level. The main role of the maintainers at the PSP level is to filter out all duplicate problem reports and only to report on unique problems to the maintenance execution level.

It is very important that the maintainers at the PSP level report on only unique software problems to the maintenance execution process level. Transferring duplicate problems implies enormous burden to the maintenance execution process level. Instead of concentrating on conducting changes to software systems, the maintenance engineers must manage duplicate problem reports. This disrupts the continuity of their work and adversity affects their productivity.

Communicating Problem Solutions to Customers

Solutions to problems should also be communicated according to a specific pattern. The status of problem resolution or the final resolution results should always be communicated to the customers via the upfront maintenance. This is the only way of avoiding chaos and a way of ensuring that all upfront maintenance organizations get informed about the progress of problem resolution.

Due to the fact that the upfront maintenance process is usually conducted by many collaborating organizations, these organizations must share knowledge and experience with each other. All important problems and their solutions should be globally recorded and shared by these organizations (Annick, 1993). By sharing this knowledge among a large body of physically distributed upfront maintenance organizations, duplicate problems may be communicated more efficiently. This would lead to a greater customer satisfaction, better maintainers' productivity, and, of course, lower maintenance costs (Annick, 1993).

Upfront Maintenance Today

Today, the domain of upfront corrective maintenance is very costly. It is estimated that to give support to the customers world-wide costs about 80% of the organization's IT related investments (Bouman, Trienekens, & van den Zwan, 1999). The main reason for this high cost is that some products may be supported by many organizations, which is mainly due to fast globalization of the markets, world-wide geographical distribution of customers, rising dependency on exports, increasing complexity of new products, and the increasing frequency of new releases (Bouman et al., 1999; Nihei, Tomizawa, Shibata, & Shimazu, 1998; Zuther, 1998). Another reason for this very high cost of upfront maintenance is the immaturity of upfront maintenance processes in comparison with other software engineering disciplines. So far, it has been given little visibility and respect within the academia or industry; hence, it has been little formalized (Bouman et al., 1999; Hart, 1993). To the knowledge of the author of this chapter, there are few models suitable for this domain (Bouman et al., 1999; Kajko-Mattsson, 2001b; Niessink, 2000).

To conduct upfront maintenance is a difficult and demanding task. The same problem in one product may appear in different environments, under different guises, and with different frequencies. To efficiently help customers, the upfront maintenance engineers must have sufficient knowledge of the products, their environments, the problems hitherto encountered in these products, and even the knowledge of customers and their businesses.

By keeping track of the customers, their products and problems, the upfront maintenance process contributes to the optimization of the product, the optimization of customer satisfaction, and optimization of the development and maintenance processes. The customer profits from the upfront maintainers' know-how, and the maintenance organization gains data on product quality, adherence to user expectations, product use patterns, and the requirements for future improvements or enhancements (Hart, 1993; Zuther, 1998).

Product Perspective

Where exactly in a software product do we make corrective changes? Let us have a look at Figure 5. We identify two phases during which the corrective changes are implemented (Kajko-Mattsson, 2001a). They are the following:

Figure 5. Corrective maintenance—product perspective (Kajko-Mattsson, 2001a).

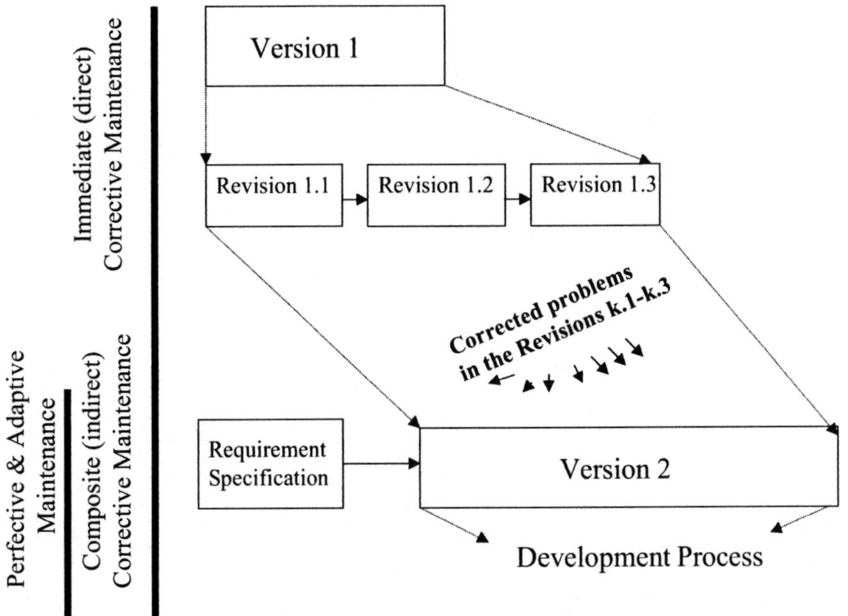

- *Immediate (direct) corrective maintenance:* All software problems are reported for a certain product version, for instance, Version 1 in Figure 5. These problems are successively attended to as they arrive at the postdelivery/ postrelease maintenance process. The correction of one particular set of problems leads to the creation of a new revision of a software product, for instance, Revision 1.1 in Figure 5. This new revision is then sent to the customers affected by the resolved software problems. It is also announced as available for use to all other customers.
- *Other maintenance and composite (indirect) corrective maintenance:* New requirements for enhancements and/or adaptations (*Other Maintenance*) are implemented in Version 2 in Figure 5. In addition, all corrections of software problems implemented in the revisions of Version k are inserted. We call the latter *Composite (Indirect) Corrective Maintenance.*

When measuring effort within corrective maintenance, we should consider both direct and indirect corrective maintenance. There are, however, some exceptions to this division. One of them concerns major requirement problems. These problems are attended to during other non-corrective maintenance categories together with new requirements for enhancements/adaptations. They should, however, still be classified and measured as corrective maintenance.

Different versions of a software product may be involved in the resolution of one. A very simplified scenario is depicted in Figure 6. Suppose now that a maintenance group is working on Version 5. This version is going to be enhanced with new functionality. The latest released version was Revision 4.2.

An external software problem has been reported in Version 3. A maintenance engineer must first recreate the reported problem. The choice of the releases in

Figure 6. Product versions involved in the problem management process.

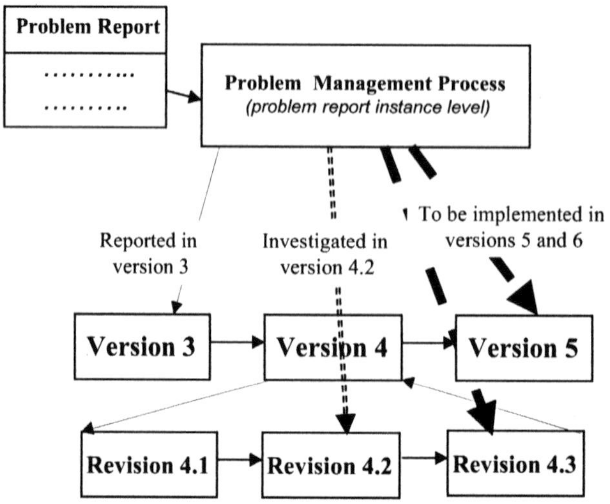

which the problem is recreated depends on many factors. Usually, the maintenance engineer attempts to do it on the version in which the problem was reported and the latest released version of the software product. In our context, it would be Version 3 and Revision 4.2. If the problem can be recreated in Revision 4.2, change actions must then be suggested and implemented. In our simplified scenario, changes would be implemented in Revision 4.3 and Version 5. It may happen, however, that the customer does not have an appropriate environment to install Revision 4.3. In this case, the maintenance organization either creates a new Revision 3.1 or advises the customer to make an environmental upgrade (Kajko-Mattsson, 2001a).

Some organizations, however, manage only one version of a software product. In that case, the choice of the version in which the problem solution will be implemented is not an issue of consideration.

Life Cycle Perspective

The roadmap presented in Figure 5 may be easily mapped on the life cycle roadmap outlined in Figure 7, in which we attempt to place corrective maintenance within the total software life cycle. We designate three main maintenance phases: (a) predelivery/prerelease phase, (b) transition phase, and (c) postdelivery/postrelease phase.

Predelivery/Prerelease Phase

According to some authors (Jones, 1994; Martin & McClure, 1983; Pigoski, 1997; Schneidewind, 1987; Swanson, 1999), postdelivery maintenance phases are greatly influenced by the level of maintainers' involvement during the predelivery stage. Irrespective of who is going to maintain the software system (either the development organization or a separate maintenance organization), there is a need for a predelivery/prerelease maintenance phase. Certain maintenance activities must be conducted then. These activities are for the most part of a preparatory nature for the transition and postdelivery/postrelease phases. Examples of these activities are the designation and creation of maintenance organization(s), their involvement during this phase, creation and realization of a maintainability plan and of a maintenance plan, preparation for the transition phase, and other things (Kajko-Mattsson, 2001e). From the corrective maintenance perspective, it is important to build maintainability into the product. The more maintainable the product, the easier it is to correct it, and the easier it is to manage the impact of change.

Transition Phase

Software transition is a controlled and coordinated sequence of activities during which a newly developed software system is transferred from the organization that conducted development to both the customer and maintenance organizations, or a modified (corrected, in our case) software system is transferred from the organization that has conducted modification to both the customer and maintenance organizations. From the corrective maintenance perspective, these activities are conducted each time an organization releases a corrected version of a software system, or a new enhanced version containing the corrections from the earlier version's revisions. Please observe that the organization that has developed new software, the organization that has modified the software, and the user organization may be one and the same organization.

Postdelivery/Postrelease Phase

Postdelivery begins after the software product has been delivered to the customer and runs all the way to the retirement activity. It includes the sum of activities to conduct changes to software systems. The changes to products are implemented incrementally in subsequent product releases (see Figure 7). Hence, postdelivery consists of a series of postrelease phases.

Usually, the creation of one major software release corresponds to the modification of software due to modification requests for either enhancements and/or adaptations. When implementing these modifications, however, we should not forget to treat this subphase as some kind of a predelivery phase for the next release. From the corrective maintenance perspective, it is very important that in this phase we preserve maintainability of the product when infusing changes. To distinguish this phase from the original development, we call it a prerelease phase.

Figure 7. CM³ Roadmap: Maintenance—software life cycle perspective (Kajko-Mattsson, 2001d).

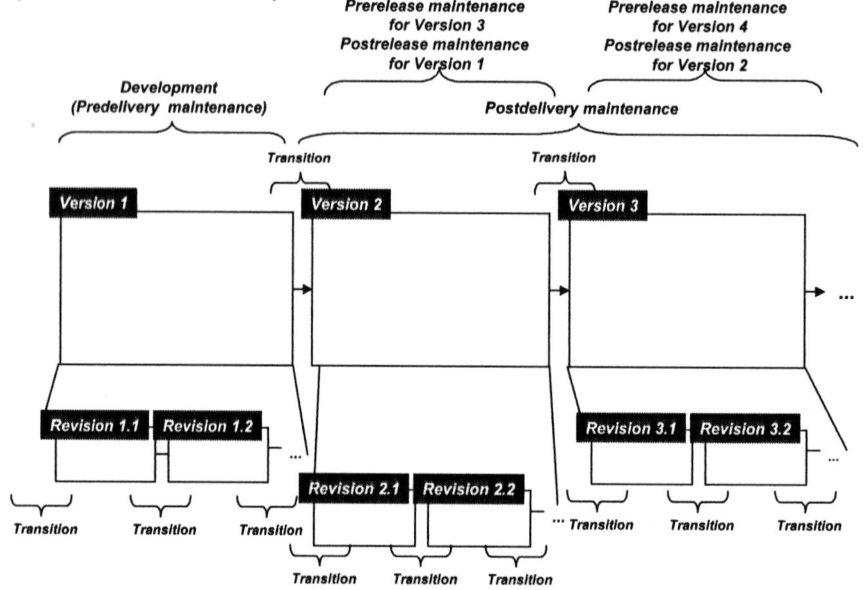

Problem Management at the Maintenance Execution Process Level

Levels of Problem Management

As noted in Figure 8, problem management process is conducted on two levels:

- *Organization-wide process level*: At this level, the problem management process is run continuously. It manages the information about all software problems in all software products within one or several organizations. The maintenance database is located at this level.
- *Individual problem report process instance level*: At this level, an instance of a problem management process is entirely dedicated to one and only one software problem.

Each software problem leads to the individual problem report process instance level. As soon as a software problem gets reported, a process instance on the individual problem report level is generated. The number of active process instances within an organization corresponds to the number of the reported software problems currently under resolution. In the context of Figure 1, we would have to run five separate processes for each individual problem report. The process at the organization-wide level is used for supervising these individual process instances and for providing different roles within the organization with different types of data on corrective maintenance process (Kajko-Mattsson, 2001a).

Roles

Various roles are involved in problem management. They are the following (Kajko-Mattsson, 2001a):

- *Problem Report Submitter (PRS)* reports on software problems. A problem submitter is either a customer organization or a maintenance engineer on any organizational process level, as depicted in Figure 3.

Figure 8. Levels of problem management.

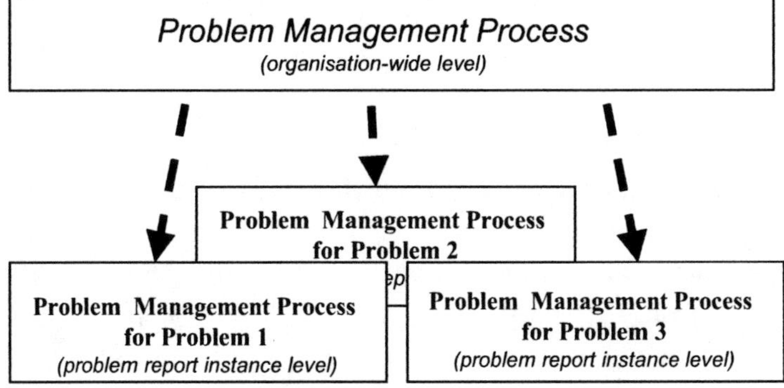

- *Problem Report Owner (PRO)* is responsible for a certain product or a product component. The *Problem Report Owner* controls and administers the problem resolution process of all problems inherent in her product/product component by supervising the work of *Problem Report Engineer(s)* (see below). The responsibility of PRO may be delegated to the *Problem Report Engineer.*
- *Problem Report Engineer (PRE)* conducts various problem management activities designated by the *Problem Report Owner*. Usually, the *Problem Report Engineer* investigates problems, identifies problem causes and root causes, designs problem solutions, and implements them.
- *Problem Report Administrator (PRA)* is responsible for the management of problem reports in general and for the smooth execution of the problem management process. The PRA is also in charge of the *Change Control Board* (see below).
- *Change Control Board (CCB)* is a group of several role holders responsible for making decisions on the management of problem reports. The composition of CCB varies across organizations. Usually, the roles inherent within the CCB are those mentioned above plus engineers from upfront maintenance (support) organizations, line managers, project coordinators, and others.
- *Problem Report Repository and Tracking System (PRR&TS)* is a tool supporting the problem management process. It records all the data about software problems and their management in its database, conducts some of the automatic problem management process steps, and provides the various role holders within the organization with useful product, resource, and process information.
- *Maintenance Team* is a group consisting of a *Problem Report Owner* and several *Problem Report Engineers*. A maintenance team is usually responsible for managing a particular component of a software system.

Problem Management Process Phases

As presented in Figure 9, a problem management process consists of three main phases, namely *Problem Reporting*, *Problem Analysis,* and *Problem Resolution*. *Problem Analysis* is then further divided into process subphases—*Report Quality Control and Maintenance Team Assignment, Problem Administration and Problem Report Engineer Assignment, Problem Investigation, Problem Cause Identification,* and *Root Cause Analysis*. *Problem Resolution* phase is divided into subphases—*Modification Design, Modification Decision,* and *Modification Implementation*. Below, we give a short introduction to these phases and their sub-phases. They are described in greater detail in the sections titled "Problem Reporting Phase" and "Problems Resolution Phase."

- *Problem Reporting Phase:* During this phase, problems get reported to the maintenance execution process level. External software problems are transferred from upfront corrective maintenance, whereas internal software problems are directly reported to this level by any role within the maintenance organization. This phase is conducted on the SPAP level of Figure 3.
- *Problem Analysis Phase*: During this phase, maintenance engineers attempt to recreate the reported problems and identify their cause(s). For some problems, they also attempt to find root causes – deficiencies in the development or maintenance processes, or deficiencies in the resources or products that have led to the reported software problem. We divide this phase into the following subphases:

  ➢ *Problem Analysis I, Report Quality Control and Maintenance Team Assignment:* In this subphase, *Problem Report Administrator* (PRA) conducts a preliminary quality control check of a problem report, and, if satisfactory, assigns the report to the relevant maintenance team. In some organizations, this subphase is automatically performed by the *Problem Report Repository and Tracking System* (Kajko-Mattsson, 1999a). Many organizations, however, follow this procedure manually. This is because it is not always easy to automatically identify the relevant maintenance team. This phase is conducted on the SPAP level of Figure 3.

  ➢ *Problem Analysis II, Problem Administration and Problem Report Engineer Assignment:* During this problem analysis subphase, the problem report has been assigned to a maintenance team and a *Problem Report Owner (PRO)*. The *Problem Report Owner* administers the reported software problem. The goal of this stage is to make a preliminary analysis of the problem report in order to determine whether it has been reported to the appropriate maintenance team, and to start planning for the problem analysis and resolution phases by analyzing the nature of the problem, by preliminarily

Figure 9. Problem management process phases.

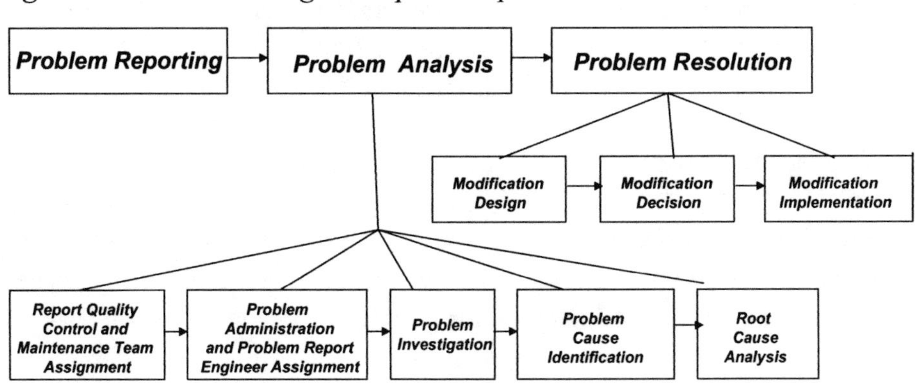

designating the release(s) to contain (a) modification(s), and by assigning the report to an appropriate maintenance engineer—*Problem Report Engineer (PRE).* This phase is conducted on the SPAP level of Figure 3.

➢ *Problem Analysis III, Problem Investigation:* During this analysis subphase, the *Problem Report Engineer* attempts to confirm the presence of the software problem. The exact steps taken during this stage may vary depending on the nature of the problem. The key focus is to reproduce the problem, mainly by executing software. At this subphase, the problem report acquires the status value *"Problem Investigated."* This phase is conducted mainly on the MDRP level of Figure 3.

➢ *Problem Analysis IV, Problem Cause Identification:* During this analysis subphase, the *Problem Report Engineer* attempts to identify underlying causes of the reported software problem, that is, defects. Defects can arise in all types of system documentation (requirement, design, test case specifications, software code, user manuals, and others). This phase is conducted mainly on the MDRP level of Figure 3.

➢ *Problem Analysis V, Root Cause Analysis:* During this analysis subphase, the *Problem Report Engineer* attempts to identify root causes of a software problem. At its earliest, *Root Cause Analysis* should be performed after the *Problem Cause Identification* subphase. It may however be performed later after the problem has been resolved. The goal is to identify process, resource, or product deficiencies leading to the reported software problem. This phase is conducted mainly on the MDRP level of Figure 3.

- *Problem Resolution Phase:* During this phase, maintenance engineers suggest one or several solution designs to the software problems, plan for the realization of these solutions, present the solutions together with the plans to *Change Control Board (CCB)*, and implement the optimal solution suggestion, which is commonly chosen by the CCB. We divide this phase into the following subphases:

➢ *Modification Design*: During this subphase, *Problem Report Engineer* designs one or several solutions to the reported software problem, and creates plans for resolving them. For each problem solution, she conducts a preliminary evaluation aiding her in ranking the solutions and pointing out the optimal one. Each solution is then either inspected or informally reviewed. This phase is conducted mainly on the MDRP level of Figure 3.

➢ *Modification Decision*: During this subphase, *Problem Report Engineer* and *Problem Report Owner* present suggestions for problem solutions and plans for their realization to the CCB. Decisions on the choice of the appropriate (optimal) solution suggestion are then commonly made by the CCB members. This phase is conducted on the SPAP level of Figure 3.

➢ *Modification Implementation*: During this subphase, appropriate changes are effected in the software system. The actual change is identified, recorded, and inspected/reviewed. This phase is conducted mainly on the MDRP level of Figure 3.

Problem Reports

All information on software problems is recorded in the maintenance database managed by *Problem Report Repository and Tracking System*. This information is communicated with the help of special reports called *Problem Reports*. These reports consist of a substantial number of forms, each dedicated to a specific process step and process role. All the forms are linked with each other providing detailed and summarized data about problem management and regular feedback to all groups concerned within the software organization. Examples of these data are problem reporting data, project management information, status of problem management process, status and quality of software products and their releases, and experience gained during corrective maintenance. In Figure 10, we present one form utilized for reporting software problems at ABB Robotics AB in Sweden. Interested readers are welcome to study other problem report forms presented in Kajko-Mattsson et al. (2000).

Maintenance organizations distinguish between external and internal problem reports:

- **External problem reports**—problems submitted externally by end-user customers.
- **Internal problem reports**—problems submitted internally by software developers, software maintainers, testers, and anybody else within a software organization.

The classification of problem reports into external and internal ones is very important. The goal is to discover as many software problems as possible before releasing software products to the customers. External problem reports are very costly. This is mainly because many upfront maintenance and customer organizations may be involved in managing them. Another factor contributing to this immense cost is the fact that certain problems may lead to changes in several releases. All these releases will have to be tested, regression tested, and delivered to customers. For these reasons, the software organizations work hard towards minimizing the number of external problem reports and maximizing the number of internal ones (Kajko-Mattsson, 2000a).

Problem Reporting Phase

During the *Problem Reporting* phase, problems get reported to the maintenance execution process level. External software problems are transferred from the upfront corrective maintenance, whereas internal software problems are directly reported to this level by any individual within a maintenance organization.

The problem submitter is obliged to provide a good and clear description of a problem. Many times, however, it may be difficult to provide a good description. This is because a problem may be described in several ways using different terms. It can be depicted in different environments and on different system levels (Wedde, Stalhane, & Nordbo, 1995).

Figure 10. Problem reporting form to be supplied by the problem report submitter at ABB ROP.

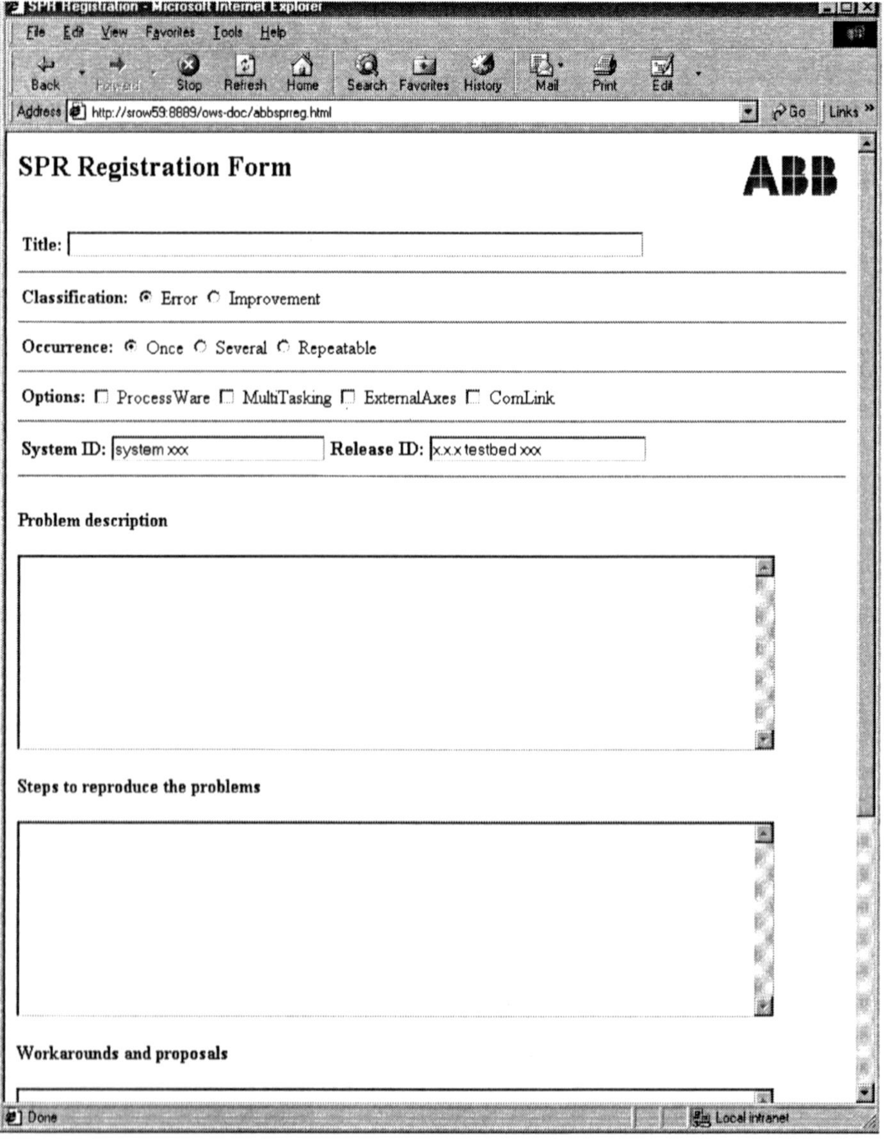

SPR Registration - Microsoft Internet Explorer

File Edit View Favorites Tools Help

Back Forward Stop Refresh Home Search Favorites History Mail Print Edit

Address http://srow59:8889/ows-doc/abbsprreg.html Go Links

SPR Registration Form ABB

Title:

Classification: ◉ Error ○ Improvement

Occurrence: ◉ Once ○ Several ○ Repeatable

Options: ☐ ProcessWare ☐ MultiTasking ☐ ExternalAxes ☐ ComLink

System ID: system xxx **Release ID:** x.x.x testbed xxx

Problem description

Steps to reproduce the problems

Workarounds and proposals

Done Local intranet

A proper problem description is the most important prerequisite for effective problem resolution. A poor, sketchy, or misleading description may lead to an enormous effort to recreate the problem, and thereby substantially retard the problem resolution process (Kajko-Mattsson et al., 2000). For this reason, the problem description should be clear, complete and correct. The problem has to be communicated in a structured and disciplined way.

To aid in minimizing the reporting time for problem submitters and in maximizing the quality of the reported data, the maintenance organization should give maximal guidance to its problem submitters on how to structure problem description data. This guidance should be built in the *Problem Report Repository and Tracking System.* It should contain templates[3] for how to describe a software problem encompassing the following (Kajko-Mattsson, 2001a):

- A general textual description of the problem.
- Description of problem effects and consequences explaining how the problem affects the software execution. This is important for assessing the severity and priority of the problem. Problems implying severe consequences should be attended to as soon as possible.
- Description of the symptoms of the problem, i.e., an observed manifestation of a problem. Compared to a consequence of a software problem, a symptom may merely indicate a minor system misbehavior. It does not always directly lead to a system failure. Symptoms can be of any kind, for instance, error messages, processing delays, blinking screens, jumping cursors, multiple cursors, misaligned text, characters doubled or omitted (Kaner, Falk, & Nguyen, 1999). Describing symptoms may greatly help maintenance engineers understand the software problem and localize its cause. Together with the information on problem consequences, it may also prove useful for recognizing duplicate problem reports (Lee, 1996).
- Description of the problem conditions such as every year shift, when particular data values are combined, or when a particular combination of software and hardware environment is executed. This information greatly helps in problem reproduction.
- Description of how to reproduce the problem providing a clear specification of steps of how to get a software program into a particular erroneous state.
- Different types of attachments containing physical evidence of the existence of the problem.

Not always can problem submitters provide enough data on software problems. In cases when little data on a problem is available and/or the problem is irreproducible, guesses are welcome if they are noted as such. Problem description should be complemented with the following data:

- Identification of the software product, its release, and components in which the problem was encountered. If a maintenance organization maintains more than one product that is being used by several customers, and if these customers possess different releases of the products, it is then imperative that the product and its release be identified (Hazeyama & Hanawa, 1999).
- Identification of the product environment describing the connection between the product and elements of its environment, such as hardware platform, software platform, development process. This piece of information is important for recreating a software problem.
- Problem type, such as software problem (requirement, design, code, documentation problem), hardware problem, operation problem. This value provides feedback for estimating the amount of corrective work.
- Submitter's judgment of problem priority and severity. This value should be considered by the maintainer when determining the maintainer's own judgment of severity and priority and when planning the maintenance work.
- Date and time of the problem occurrence. This value is relevant for only some systems, for instance, the safety-critical systems and the systems in which time is a critical performance factor. Its value in combination with the severity value, enables the assessment and prediction of the reliability of the software system. It may also be needed to reproduce or analyze problems that are keyed to the time of day such as workloads.
- Problem submitter may describe software problems more thoroughly than this. For more information, interested readers are welcome to review Kajko-Mattsson (2001a).

Problem Analysis I: Problem Quality Control and Maintenance Team Assignment

In this phase, the *Problem Report Administrator* conducts a preliminary quality check of the contents of a problem report. She checks whether the problem report is legible, whether the description is exhaustive enough, whether it appropriately depicts the problem, and whether the product is properly identified. If the problem report is not satisfactory, the *Problem Report Administrator* contacts the problem submitter for clarification and supplementary information.

A problem report that is incorrect, incomplete, or inconsistent carries varying degrees of risks (Arthur, 1988; Kajko-Mattsson, 2001d). It leads to inappropriate understanding of the software problem. Misunderstood problems severely obstruct the problem resolution process. More work has to be done on the communication about the problem with upfront maintenance and customers, on its investigation, and resolution. This, in turn, substantially affects maintainers' productivity. An incorrect description, especially of externally reported problems, is one of the main reasons for the very high cost of the problem management. At its worst, it may lead to several months of hard labor for the maintainer to recreate the problem.

If the problem report is satisfactory, the *Problem Report Administrator* assigns it to an appropriate maintenance team. It may happen that the problem is so comprehensive so that it must be attended to by several collaborating teams, or even organizations. The identification of the maintenance team and/or organization is not always so straightforward when reporting on external software problems. Some systems are very complex. They may be embedded systems, integrated with other systems, and their subsystems may not always be produced by the same organization. In such a case, it is not always easy to identify the affected subsystem and its release at the problem encounter. The organizations possessing such subsystems must have a special procedure for system identification and maintenance team assignment. Usually, representatives from different systems or system parts have a meeting or a series of meetings during which they attempt to analyze the problem and localize it within such a complex system.

Problem Analysis II: Problem Administration and Problem Report Engineer Assignment

All problem reports transferred by the *Problem Report Administrator* are automatically reported to the *Problem Report Owner,* the manager of a maintenance team. The *Problem Report Owner* administers the reported software problem and makes a preliminary analysis in order to determine whether it has been reported to the appropriate maintenance team. She also checks the quality of the reported data; that is, she repeats some of the steps conducted by the *Problem Report Administrator*. It may happen that she revises the problem description or sends it back to the problem submitter for additional information. Even if the quality of the problem report has been checked by the *Problem Report Administrator*, the PRO may still notice shortcomings. Please observe that *Problem Report Administrator* only possesses the general knowledge of the software product.

After having checked the quality (correctness, consistency, and completeness) of the problem descriptions, the *Problem Report Owner* attempts to verify the problem. At this stage, she only verifies that the problem report communicates a problem. If not, she rejects the problem report and informs the upfront maintenance personnel that the reported problem was not a problem at all.

If the report communicates a software problem, then the *Problem Report Owner* attempts to create a preliminary mental picture of the problem and of the problem solution. She must do this in order to build an opinion about the problem, preliminarily evaluate it and its presumptive solution, and decide whether it is worth proceeding with its resolution. If so, then she must make preliminary plans for implementing problem solutions. Due to the fact that little is known about the problem at this stage, the plan for implementing the problem solution is very coarse-grained. It merely consists of the designation of the software version(s) in which the problem will be investigated and possibly implemented, and the designation of an activity, such as *Problem Investigation*. When doing this, she assigns the problem report to one of her team members, the *Problem Report Engineer*.

Problem Analysis III: Problem Investigation

In this analysis subphase, the *Problem Report Engineer* attempts to confirm the presence of the software problem. Before doing so, however, she must study the problem report and check its correctness, consistency, and completeness. She must do so even if this step has already been conducted by both the *Problem Report Administrator* and *Problem Report Owner*. Many organizations today impose such a continuous quality control and analysis of the problem report data. The worst mistake the maintenance engineers can make is to begin the problem resolution process without thoroughly analyzing the reported problem (Myers, 1976). Not paying enough attention to the problem description, may lead maintenance engineers to (1) misunderstand the problem or mistake it for some other one, (2) underestimate the resources required for its resolution, and/or (3) make too optimistic a schedule for its resolution (Arthur, 1988).

The *Problem Report Engineer* may still notice shortcomings in problem descriptions at this stage. She may either revise them or send them back to the *Problem Submitter*. Due to the fact that the understanding of the nature and complexity of a problem becomes clearer as problem resolution progresses, problem analysis should continue throughout the problem resolution process. Its results should be continuously revised and recorded. This is because the more we work with the problem, the better understanding we gain about it.

After the above-mentioned analysis and control, the *Problem Report Engineer* investigates the problem. Problem investigation may be one of the most time-consuming activities within corrective maintenance. Usually, it takes between half a day to one day to confirm a problem. At its worst, however, if a problem is very complex, it may take several months of hard labor to recreate a problem in one version of a product.

An investigation of one problem is conducted both at the upfront and maintenance execution process levels (see Figure 3). At the upfront maintenance level, the maintainers are obliged to confirm the problem by recreating it before reporting it to the next maintenance level. At the maintenance execution process level, the maintainers should, however, conduct their own problem investigation by following the results and guidelines from the upfront maintenance process level. They may also have to investigate several releases. Figure 11 shows en excerpt from *Corrective Maintenance Maturity Model (CM^3): Problem Management* delineating the process steps to be conducted by these two maintenance levels (Kajko-Mattsson, 2001a).

After problem investigation, the *Problem Report Engineer* has a more or less clear picture of the software problem. However, this picture is not clear enough for designing a problem solution. Just as the *Problem Report Owner* does, the PRE attempts to create or refine a preliminary mental picture of the problem solution. She also refines its preliminary evaluation and the preliminary plans. In these plans, she suggests *Problem Cause Identification Activity* and identifies the versions in which the problem causes (defects) should be identified. She reports all this to the *Problem Report Owner* who in turn makes decisions whether to continue with the problem management.

Problem Analysis IV: Problem Cause Identification

After the software problem has been confirmed, the *Problem Report Engineer* attempts to identify and localize its cause. Usually, the engineer who has conducted problem investigation continues with problem cause identification. It may happen, however, that another engineer may have to take over the problem management. Irrespective of whether it is the same *Problem Report Engineer* or not, the first step to be conducted during this phase is to study the results of the

Figure 11. An excerpt from CM³: Problem Management *(Kajko-Mattsson, 2001a).*

ME-Process-PI-3.1.1:	***CM³: Upfront Maintenance: Problem Investigation:*** ←	***Upfront maintenance process level***
ME-Process-PI-3.1.1.1:	Study the problem report.	
ME-Process-PI-3.1.1.2:	Identify the uniqueness of the problem.	
ME-Process-PI-3.1.1.2.1:	Identify and classify the symptoms of the reported problem.	
ME-Process-PI-3.1.1.2.2:	Use a list of known problems in order to determine the uniqueness of the software.	
ME-Process-PI-3.1.1.3:	If the problem report is a duplicate:	
ME-Process-PI-3.1.1.3.1:	Check the progress of the problem resolution.	
ME-Process-PI-3.1.1.3.2:	Update the master problem report with additional information on the problem found in the duplicate problem report, if any.	
ME-Process-PI-3.1.1.3.3:	Link the duplicate problem report to its unique correspondence.	
ME-Process-PI-3.1.1.3.4:	Close the management of the duplicate software problem report.	
ME-Process-PI-3.1.1.3.5:	Report to the customer on the status of the problem resolution.	
ME-Process-PI-3.1.1.4:	If the problem is not recognised as a duplicate:	
ME-Process-PI-3.1.1.4.1:	Recreate the problem.	
ME-Process-PI-3.1.1.4.2:	Record the results of the problem investigation.	
ME-Process-PI-3.1.1.4.3:	Deliver the results to the maintenance execution process level.	***Maintenance execution process level***
ME-Process-PI-3.1.2:	***CM³: Problem Management: Problem Investigation:*** ←	
ME-Process-PI-3.1.2.1:	Check the results of problem investigation at the upfront maintenance level.	
ME-Process-PI-3.1.2.2:	Irrespective of whether the problem has been successfully recreated by CM³: Upfront Maintenance, recreate the problem anew by conducting the following steps:	
ME-Process-PI-3.1.2.2.1:	Study the problem report.	
ME-Process-PI-3.1.2.2.2:	Designate (an) appropriate version(s) of the product in which the problem will be investigated.	
ME-Process-PI-3.1.2.2.3:	Study the software system.	
ME-Process-PI-3.1.2.2.4:	Create a pertinent execution environment for the version(s) of the product in which the problem will be investigated.	
ME-Process-PI-3.1.2.2.5:	Define a set of test cases required for problem investigation.	
ME-Process-PI-3.1.2.2.6:	Execute the software system to recreate the problem.	

problem investigation in order to get acquainted with (if another engineer) or reacquainted with the problem. Afterwards, the engineer starts looking for problem causes. For each software system documentation level (requirement, design, software code, test and regression test specifications, customer documents, and others), she studies its documentation items and checks their correctness. If the problem cause(s) (defect(s)) is/are found, she records them. However, she should not conduct any changes yet.

After having identified problem causes, the *Problem Report Engineer* has a clearer mental picture of the problem solution. However, this picture is still hazy. At this stage, she refines the preliminary solution and the preliminary plans. Finally, she reports all this to the *Problem Report Owner* who in turn makes the decision whether to continue with the problem management.

Problem Analysis V: Root Cause Analysis

After having identified defects, it is possible to start identifying process, product, or resource deficiencies leading to the reported software problem. This is done during root cause analysis. At its earliest, root cause analysis may be conducted after problem causes have been identified. Many organizations do so, for instance, ABB Robotics AB. There are, however, some organizations, for instance, Ericsson Radio Systems, who conduct root cause analysis at the end of problem resolution phase. During this phase, the *Problem Report Engineer* attempts to identify the process activities during which the problem cause was introduced. She analyzes these activities and attempts to identify and classify the root causes of the problem. Examples of classes of root causes may be vague requirements in the specification phase, lack of product knowledge, lack of process knowledge, weakness in specification of function in the specification phase, weakness of testing procedure in the test phase, wrong CCB decision and others (Kajko-Mattsson et al., 2000).

Very few software organizations conduct root cause analysis today (Kajko-Mattsson, 2001a). These organizations do not do it on every software problem. This would be too expensive. The candidates for root cause analysis are either all external software problems, very serious internal and/or external ones, or groups of very frequently encountered problems.

Problem Resolution: Modification Design

After the problem cause(s) has/have been identified, the *Problem Report Engineer* must design one or several problem solutions to the reported software problem. Figure 12 presents an excerpt of the CM^3 process steps to be conducted during this phase. The first task to be conducted during this phase is to study the results of the problem investigation and problem cause identification in order to get acquainted with (if done by another engineer) or reacquainted with the software problem and its causes. After that the engineer may start designing problem solutions. A common mistake is that a maintenance engineer spends substantial time on investigating a problem and its causes, and then immediately starts making changes to the software system without taking time to design a solution. Even if the problem does not appear complex and difficult to resolve, it still requires a design. A good design helps in the implementation of change and in preserving the maintainability of the software.

When designing problem solutions, the engineer must verify the choice of the software releases in which these solutions will be implemented. As we have already mentioned, one and the same problem may have to be resolved in several releases. The maintainer must determine firm requirements for the change, that is, identify where exactly these changes should be infused. She must also study the history of the changes to the components now earmarked for change, so that the suggested changes are compatible with the earlier changes. If the *Problem Report Engineer* makes several modification designs (problem solutions), she must evaluate and rank each design for its optimality with respect to short-term and long-term benefits, drawbacks, costs, risks, and the like.

For each modification design, the *Problem Report Engineer* creates a plan for its realization. Such a plan contains (1) identification of the version(s) of the software product in which the software problem will be resolved, (2) designation of the activities to be taken, (3) identification of all the resources required for implementing the problem solution (equipment, personnel resources), (4) identification of the prerequisites for implementing the problem solution, and (5) determination of the schedule for conducting the activities/tasks.

Figure 12. An excerpt from CM³: Problem Management: Modification Design *(Kajko-Mattsson, 2001a).*

ME-Process-MD-5.1:	Mark out of the most important issues to be covered by a process model.
ME-Process-MD-5.1.1:	Conduct a mini version of problem investigation process.
ME-Process-MD-5.1.2:	Conduct a mini version of cause identification process.
ME-Process-MD-5.1.3:	Design at least one problem solution.
ME-Process-MD-5.1.3.1:	Verify the choice of the software release(s) to contain the problem solution.
ME-Process-MD-5.1.3.2:	Design a problem solution. For all software system documentation level items identified during *Problem Analysis IV: Problem Cause Identification* (requirement and design specifications, software code, test and regression test specifications, customer documents, and other), do the following:
ME-Process-MD-5.1.3.2.1:	Determine firm requirements for the change.
ME-Process-MD-5.1.3.2.2:	Consider the history of prior changes to the documentation item(s), both successful and unsuccessful. The goal is to find out whether the change is compatible with earlier changes.
ME-Process-MD-5.1.3.2.3:	Design (a) change(s) to the documentation item (revise the software items identified during *Problem Cause Identification*).
ME-Process-MD-5.1.4:	Rank each suggestion for its optimality with respect to short-term and long-term benefits, drawbacks, costs, risks, design simplicity, complexity of the components to be changed, change scope, testability, maintainability, usability across different software versions and other matters.
ME-Process-MD-5.1.5:	If the problem solution is designed earlier than during *Modification Design*, the maintainer should at least create a mental picture of what a problem suggestion would look like. This mental picture should be based on the steps defined in ME-Process-MD-5.1.3.2. This mental picture, when documented, becomes the skeleton of a problem solution.

The modification designs should be either reviewed or inspected. Usually, major problems are formally inspected and minor problems are only informally reviewed. It may also happen that suggestions for solutions of minor problems are not reviewed at all. We are of the opinion, however, that each suggestion for change should be inspected/reviewed.

Before presenting the modification design to the *Change Control Board*, the *Problem Report Engineer* presents the results to *Problem Report Owner*, who in turn makes the decision whether to present the modification designs to the CCB.

Problem Resolution: Modification Decision

Software products are assets of the organization, not of individual engineers. It is not enough that problem solutions and plans for their realization are approved by the engineers' closest manager (*Problem Report Owner*). There must be an organization-wide authority for approving the appropriate suggestion for problem resolution and for approving a plan for its realization. Such a group is usually called a *Change Control Board (CCB).* Its members should represent various roles within an organization who have interest in deciding upon the change. It is usually composed of members representing different functional areas, such as high-level managers, *Problem Report Owners*, *Problem Report Engineers,* upfront maintainers, and sometimes even customers.

During the CCB meetings, the maintenance engineer (either *Problem Report Owner* or *Problem Report Engineer*) presents the suggestion(s) and plan(s) for its/their realization to the CCB. The CCB members make an evaluation of the problem solution(s). They analyze the plan(s), and make decisions on each suggestion. These decisions may be the following: (1) the suggestion is accepted and is going to be implemented as planned, (2) the suggestion is accepted but plans must be modified, (3) the suggestion is rejected, (4) the suggestion is accepted, but its solution is deferred, or (5) the suggestion is accepted, but must be modified due to the issues raised during the CCB meeting. All the CCB decisions must be documented.

CCB should be officially responsible for all changes to the software and for its allocated budget. Its task is to ensure that the implementation of problem solutions meets the long-term goals of the organization (Briand, Basili, Yong-Mi, & Squier, 1994). CCB meetings give managers more insight into the corrective maintenance process and allows them to exchange relevant information about the changes to the software systems. They can help eliminate possible bias towards some solutions and to resolve conflicting change requests (Martin & McClure, 1983), for example, in cases when one maintenance team may wish to implement a problem solution that will hamper the implementation of another modification to be implemented by some other maintenance team. CCB meetings are also an excellent means for keeping up discipline within a maintenance organization. Forcing maintenance engineers to regularly report on the progress of their problems ensures that productivity is kept at a satisfactory level.

There has been some criticism that the CCB slows down the process. Usually, CCB meetings take place once a week or a fortnight (Kajko-Mattsson, 1999a and 1999b). This may force engineers to wait with the implementation of even minor obvious changes. For this reason, many organizations allow their maintenance engineers to start resolving minor problems without waiting for the CCB decision. Their suggestions, even if they are resolved, should still be discussed at the CCB meetings. There is has the risk however, that the maintainer's effort spent on implementing a change may be lost if the CCB rejects the solution.

Problem Resolution: Modification Implementation

During *Modification Implementation,* the *Problem Report Engineer* implements the changes to the software system according to the modification design (problem solution) that has been chosen by the CCB. When implementing the problem solution, the maintenance engineer does not only change software code and documentation, but also makes additions, deletions, and rewritings. Last but not least, the maintenance engineer must determine the impact of these modifications on other parts of the system. This may involve a profound study of the software system. The maintenance engineer must make certain that the new logic is correct, the unmodified portions of the system remain intact, the system still functions correctly, consistency is preserved, maintainability is not reduced, and customers are not adversely affected (Martin & McClure, 1983).

Not considering the above-mentioned issues may lead to an ineffective change implementation, which is the main cause for product aging. To avoid this, maintenance engineers should conduct changes in a formal and tractable way. Figure 13 presents an excerpt of the CM^3 process steps to be conducted during this phase.

During modification implementation, the *Problem Report Engineer* should study the results of the *Problem Investigation* and *Cause Identification* phases. She should also make sure that she understands the suggestion for the modification and the software to be modified. After infusing changes, she should test the changed software units (unit testing).

The modification should be conducted according to the documentation and code standards imposed by the maintenance organization. The engineer should follow these standards. In this way, the organization may preserve the maintainability of the software system. All changed components should be identified on as low a granularity level as possible. This information should be visible via the problem report so that relevant feedback may be provided for the cost assessment of corrective maintenance. Finally, the engineer should check the consistency of all changed components with those unchanged ones to ensure the traceability of the system, and she should check whether all changes in software are traceable from problem report and vice versa to ensure the traceability of change.

Testing

During problem resolution, the *Problem Report Engineer* has implemented changes to the affected software components. She has not, however, thoroughly tested these changes. Usually, she conducts unit testing and/or component testing (testing of a set of strongly related software and/or hardware units). During these tests, she repeats the test cases that have been conducted during the problem investigation for recreating the software problem. She may also design new complementary test cases to cover the additional changes made during modification design. Finally, she must conduct regression testing, that is, repeat the suite of all the former tests for those units. She must do it to make sure that the new changes have not affected other parts of the software component. Before regression testing, however, she should clean up the regression test repository. It may happen that the new changes have made the old test cases obsolete.

The integration tests are usually conducted by the departments responsible for the product components, whereas the system tests are conducted by the separate testers within the organization. Unfortunately, many maintenance organizations do not system test the releases that are being the result of corrective changes. But if they do, then testing teams tightly cooperate with the upfront maintainers. It is the upfront maintainers who know best how their customers use the product. Hence, they are an important resource for testing the adherence to user expectations, product use patterns, problems, and other matters.

Figure 13. An excerpt from CM^3: Problem Management: Modification Implementation *(Kajko-Mattsson, 2001a).*

ME-Process-MI-3.1.1:	Conduct a mini version of the *Problem Investigation* process. Make sure that you understand the problem.
ME-Process-MI-3.1.2:	Conduct a mini version of the *Cause Identification* process. Make sure that all causes are identified and understood.
ME-Process-MI-3.1.3:	Study the software system. Make sure that you understand it.
ME-Process-MI-3.1.4:	Study the chosen modification suggestion.
ME-Process-MI-3.1.5:	Make the changes. For each component (documentation item) to be modified (at any software system documentation level), do the following:
ME-Process-MI-3.1.5.1:	Make the necessary change(s).
ME-Process-MI-3.1.5.2:	Test the change(s) (unit testing).
ME-Process-MI-3.1.5.3:	Check that the component is changed according to the documentation/coding standards as defined by the organisation.
ME-Process-MI-3.1.5.4:	Identify and record the parts of the component that have been changed down to the line/figure level.
ME-Process-MI-3.1.5.5:	Record the reason for the change (corrective, non-corrective change).
ME-Process-MI-3.1.5.6:	Ensure the traceability of the changed documentation item to the other (modified and unmodified) documentation items.
ME-Process-MI-3.1.5.7:	Ensure the traceability of change from the modified documentation item to the problem report and vice versa.

Process Flexibility

So far, we have described the main phases of problem management process as suggested by *CM³: Problem Management* (Kajko-Mattsson, 2001a). The choice of these phases and their subphases may, however, vary for each problem management process instance. Below, we illustrate some of the process variances:

- *Problem Investigation* may occur from zero to several times. In the first case (zero times), the *Problem Report Submitter* is a maintenance engineer who has encountered the problem. She already knows the problem. Therefore, for some less serious problems, she may directly start conducting *Problem Cause Identification* activity. In the second case (several times), the *Problem Report Engineer* must check whether the problem exists in several releases. This may be relevant in cases when different customers use different releases of the product. It is then important to identify all those releases and check whether they contain the reported problem.
- *Problem Cause Identification* may occur from zero to several times. In the first case (zero times), the *Problem Report Engineer (PRE)* might have been able to localize a minor problem cause by herself when investigating some other problem. In this case, this PRE becomes a *Problem Report Submitter*. It is also she who later attends to this problem. The *Problem Report Submitter* might also be a tester or another maintenance engineer who has encountered the problem during testing or attending to some other problem. Her duty is to give a thorough report on what the problem is and identify the problem cause(s), if possible. In cases where the problem cause has been identified by somebody else, the maintenance engineer attending to the problem should not rely on somebody else's data. She should identify the problem cause(s) by herself before designing any problem solutions. It may also happen that the *Problem Report Engineer* must identify all the possible releases in which the problem and its cause(s) have been identified.
- The process phases such as *Problem Investigation, Problem Cause Identification, Modification Design,* and *Modification Implementation* may merge into one and the same phase. This is the case when the maintenance engineers attend to some minor (cosmetic) problems. Instead of going through all these phases, they may directly implement them as soon as they have identified them. We recommend, however, that these problems and their solution designs be presented to the CCB, irrespective of their triviality. One should not forget to evaluate the impact of change as well. The presentation to the CCB may occur after the problem has been resolved. However, if the CCB rejects some solution, then the changes must be unmade.
- One problem may be attended to by many teams. Still, however, we suggest that one team should be responsible for the management of the whole problem.

- *Modification Decision* (CCB decision) may take place earlier than after *Modification Design.* Some problems are very serious and/or urgent. Hence, they must be supervised by the CCB or some other authority within the organization immediately after being reported. They may also be under continuous supervision during the whole problem management process (Kajko-Mattsson, 2001a).
- The attendance to the software problem may terminate anywhere within the problem management process. Below, we identify some of the scenarios during which the problem management may be terminated.
  - During the first *Problem Analysis* subphase, *Report Quality Control and Maintenance Team Assignment,* when checking the correctness of the reported data, the *Problem Report Administrator* may discover that the reported problem was not a problem at all; it was a misunderstanding.
  - The problem cannot be recreated. This can easily arise in the case of a distributed system or a real-time embedded system. The maintenance organization may temporarily postpones its resolution until more information is acquired about it. The customer must agree to live with the problem.
  - It is too expensive to attend to the problem, and the problem may have to be lived with and/or worked around.
  - It is too risky to resolve the problem. For instance, it may require substantial reorganization of the system, retraining of all customers, too many changes in the customer documentation, and/or operational changes.

Corrective Maintenance Maturity Model (CM^3): Problem Management

The description of problem management in this chapter is an extract of *Corrective Maintenance Maturity Model (CM^3): Problem Management* – a detailed problem management process model specialized to corrective maintenance category only (Kajko-Mattsson, 2001a). It is a constituent process model belonging to Corrective Maintenance Maturity Model (CM^3), presently under development. CM^3 encompasses the most important process models utilized within corrective maintenance. They are:

- *CM^3: Predelivery/Prerelease:* Activities to be conducted in order to prepare the software system for future corrective maintenance (Kajko-Mattsson, Grimlund, Glassbrook, & Nordin, 2001).
- *CM^3: Transition:* Activities required for a smooth transfer of a software system from development to maintenance, and activities required for the delivery of a software system from a development/maintenance organization to the customer organization(s).

- *CM^3: Problem Management:* Activities required for effective reporting, analyzing and resolving of software problems (Kajko-Mattsson, 2001a and 2001e).
- *CM^3: Testing:* Activities required to test corrective changes in a software product.
- *CM^3: Documentation:* Activities required to create a consistent and exhaustive documentation of the software system and its corrective changes (Kajko-Mattsson, 2001g).
- *CM^3: Upfront Maintenance:* Activities required for communicating with customers on software problems (Kajko-Mattsson, 2001b).
- *CM^3: Maintainers' Education and Training:* Activities required for continuous education and training of maintenance engineers (Kajko-Mattsson, Forssander, & Olsson, 2001).

Structure of CM^3

As depicted in Figure 14, each constituent CM^3 process model is based on *CM^3: Definition of Maintenance* and *Corrective Maintenance*. In addition, each such process model has the following structure: (1) *CM^3: Taxonomy of Activities* listing a set of activities relevant for the process; (2) *CM^3: Conceptual Model* defining concepts carrying information about the process; (3) *CM^3: Maintenance Elements* explaining and motivating the maintenance process activities; (4) *CM^3: Roles* of individuals executing the process; (5) *CM^3: Process Phases* structuring the *CM^3* processes into several process phases; (6) *CM^3: Maturity Levels* structuring the *CM^3* constituent processes into three maturity levels (*Initial*, *Defined*, and *Optimal*); and (7) *CM^3: Roadmaps* aiding in the navigation through the *CM^3* processes.

Figure 14. Structure of each CM^3 constituent process model.

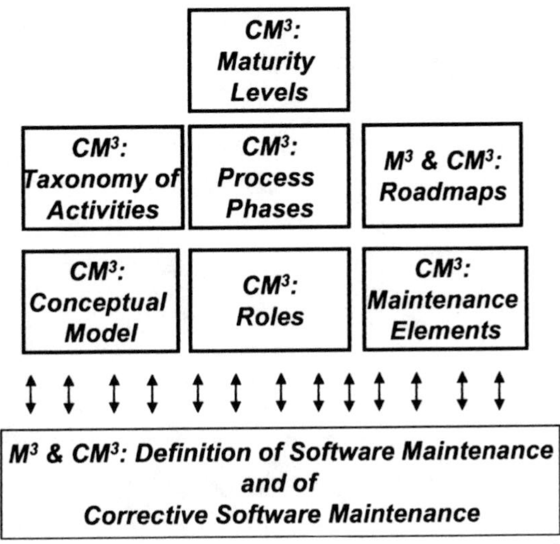

With this structure, we aim towards providing maximal visibility into corrective maintenance and towards decreasing the subjectivity in the understanding and measurement of corrective maintenance processes. We hope that the *CM^3* model will constitute a framework for researchers and industrial organizations for building process and measurement models, and serve as a pedagogical tool for universities and industrial organizations in the education of their students/engineers within the area of corrective maintenance.

The problem management process is very complex in itself. It may not always be easy for maintenance organizations to know which process issues to implement first. For this reason, we have divided our *CM^3* problem management process model into three levels, where each level provides a relevant guidance on what to implement. Below, we briefly describe our levels.

CM^3: Problem Management: Level 1 (Initial)

At the *Initial* level, the process is implicitly understood by practitioners and their managers. It offers little visibility or gives a distorted view of its status. At the very most, the organization has insight into the number of software problems reported and resolved, as may be reflected in Figure 15. But, it cannot monitor the problem resolution process, and make reliable process analyses. The feedback supplied by the process is not always meaningful.

CM^3: Problem Management identifies a very limited set of process steps and maintenance elements for this maturity level. These process steps merely indicate that problems have been reported and resolved. *CM^3: Problem Management* also strongly recommends consideration of the most rudimentary process requirements—*CM^3: Problem Management: Maintenance Elements*—as a starting point on the voyage towards process improvement.

CM^3: Problem Management recommends that all software problems, just as any maintenance requirement, be communicated in writing, not orally. Otherwise, we cannot carry out problem resolution in a formal, disciplined, and tractable way. Communication in writing is a way of achieving control over all the maintenance requirements (problem reports in our context). To avoid confusion with respect to the amount of maintenance work, one problem should be reported in one and only one problem report. This is also the most important prerequisite for enabling problem validation and verification, and an essential condition for checking that the process is being followed.

Figure 15. Problem management process maturity—Level 1.

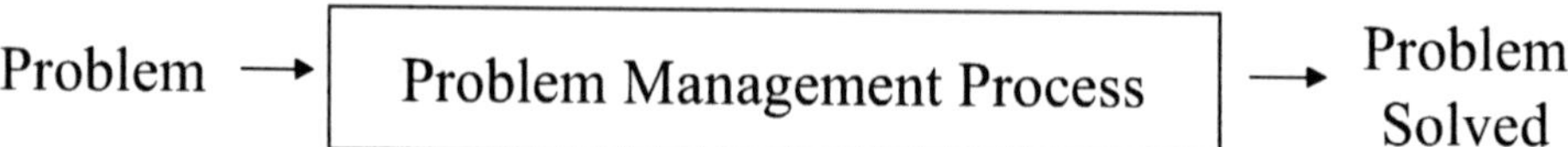

Problem submitters must be identified to enable the communication of the reported problem and the delivery of the problem solution to the right customer. To help management find out how maintenance time and effort is spent, all maintenance requests should be classified as either belonging to corrective or other (non-corrective) maintenance. This requirement also enables prioritization of the maintenance requests, and provides a basis for different kinds of process and product analyses.

To build an opinion about customer satisfaction and to be able to assess the quality of the product from the customer perspective, the organizations should distinguish between problems encountered internally and externally. At this level, the maintenance organizations should also start filtering out duplicate problem reports. Inability to do this substantially affects the maintainers' productivity and the interpretation of different types of statistics.

CM³: Problem Management: Level 2 (Defined)

At the *Defined* level, a simple coarse-grained process model for managing software problems is defined and documented (see Figure 16). The model covers the most rudimentary process phases and activities. These phases and activities offer visible milestones for following the progress and for making different kinds of intermediate decisions.

Although simple, the process is consistently adhered to. Process compliance with its documentation is checked on a regular or event-driven basis. The most rudimentary data and measurements supplied at this level appropriately mirror the process, and hence, are meaningful. Due to the coarse-grained visibility into the process, the assessment of predicted and actual effort and resources still allows only rough estimation.

Because a proper problem characterization is the most important prerequisite for efficient problem resolution, the maintenance organization should provide maximal support to problem submitters on how to describe their maintenance

Figure 16. Problem management process maturity—Level 2.

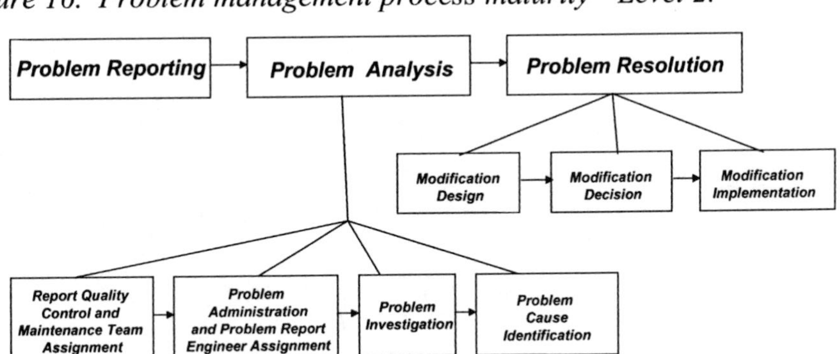

requests (problem reports in our context) by institutionalizing a template for problem description. Correctness, consistency, and completeness of the problem report data is continuously checked and improved throughout the problem management process in order to maximize an objective understanding of a software problem and to provide correct, consistent, and complete feedback for making different kinds of simple statistics and process analyses. Sources (submitter and maintainer) of problem description and problem data are separated from each other to enable efficient problem resolution, validation, and verification, to provide correct and reliable statistics, and to correctly plan the maintenance work. The problem resolution process, its major phases, results, and executing roles are documented. Process variances (allowed process deviations – still coarse-grained at this level) are defined to enable process flexibility and to provide feedback for further process refinement and improvement. Maintenance work is planned and controlled with predetermined scheduled releases. Treating the software product as an organizational asset, all changes to it are commonly approved by a *Change Control Board.*

CM³: Problem Management: Level 3 (Optimal)

At the *Optimal* level, *CM³: Problem Management* allows fine-grained visibility into the progress of problem resolution. We have clear insight into every process step and its results (see Figure 17). We may identify all the changes made to the software product on a fine-grained product level (line of source code and documentation). This in turn aids in achieving full traceability of change. The detailed process knowledge helps us analyze meaningfully the development and maintenance processes and identify their weaknesses and strengths. Classification of

Figure 17. Problem management process maturity—Level 3.

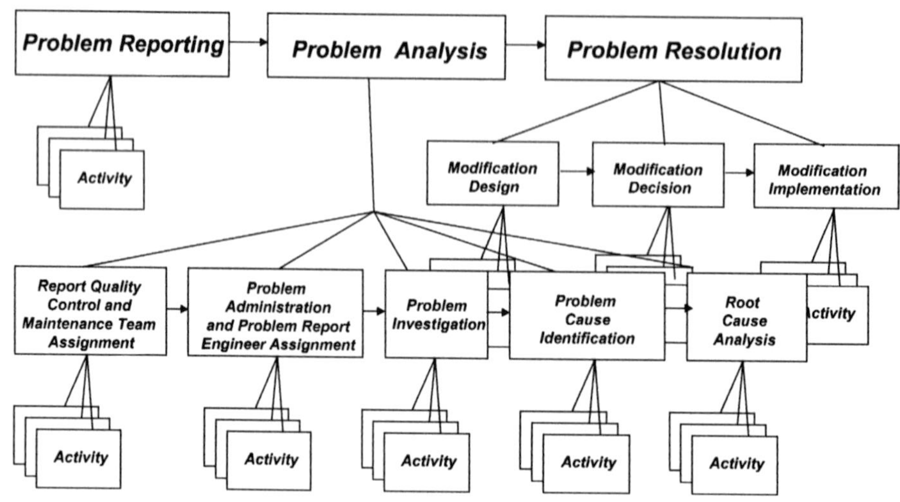

symptoms enables an automatic identification of duplicate problems, thus substantially improving the maintainers' productivity. Classification of causes helps in predicting and measuring the cost of corrective maintenance, in assessing the product quality, and in performing root cause analyses. Root cause analyses are systematically conducted, providing continuous feedback to process improvement and preventive maintenance processes.

For major activities within the problem management process, such as *Problem Investigation, Problem Analysis, Modification Implementation,* and other activities, organizations provide detailed process models of how they should be conducted. This is a way of assuring that each process activity is conducted in as uniform a manner as possible, and for assuring the quality of maintenance work.

No change is implemented without first making a thorough evaluation during which the complexity and ramifications of a software problem are determined. This substantially helps to assess the amount of effort and resources required for its resolution. Ripple effect is managed providing valuable feedback to regression testing. All suggestions for change and all change implementations are reviewed and/or inspected. In contrast to Level 2, the discrepancy between the planned and actual effort is strongly reduced due to the more detailed process feedback. The process does not only suffice for managing software problems, but also provides useful feedback to process improvement and defect prevention.

Epilogue

In this chapter, we have presented a problem management process utilized within corrective maintenance. The theory presented here has been extracted from a recently defended PhD thesis titled *Corrective Maintenance Maturity Model: Problem Management* suggesting a detailed problem management process model (Kajko-Mattsson, 2001a). By concentrating our efforts on a limited domain, we were able to scrutinize it meticulously, establish its definition, scope, borders to other processes, and, most importantly, to identify all the important process activities necessary for efficient management of software problems.

CM³: Problem Management has been built by *Software Maintenance Laboratory*, a cooperation of Stockholm University/Royal Institute of Technology in Stockholm and ABB in Västerås, Sweden. *CM³: Problem Management* has been primarily developed in the ABB context. However, it is targeted to all software organizations involved in building or improving their corrective maintenance processes. It has been evaluated against 28 industrial non-ABB processes. The evaluation results have shown that our model is realistic, down-to earth, and that it appropriately reflects the current industrial reality. *CM³: Problem Management* does the following:

- Provides a fine-grained process model allowing maximal visibility into the problem management process.
- Suggests an evolutionary improvement path for software organizations from an ad hoc immature process to a mature and disciplined one by structuring the problem management process into three maturity levels. These levels indicate the degree of the organizations' visibility into the process and the organizations' capability to manage problems and to utilize the knowledge so gained in defect prevention and process improvement efforts.
- Provides a framework within which industrial organizations can examine their practices and compare notes. It also provides students and researchers with a real sense of the size of the corrective maintenance process.
- Serves as a pedagogical tool for universities and industrial organizations in the education of students/engineers within the area of corrective maintenance. This is achieved by offering our thorough explanations and motivations for different process steps.

The problem management process presented in this chapter is only one of very many processes required for carrying out engineering chores. We should develop similar models for all other processes laying out in detail the work of developing and maintaining software systems. It is only in this way that we may achieve full insight into the development and maintenance processes, decrease our subjectivity in understanding and measuring them, increase our productivity, and contribute to the improvement of our common software engineering work overall.

ENDNOTES

1 It is expressed in the Entity-Relationship Model (Powersoft, 1997). See the Appendix, an explanation of the modelling constructs.

2 The remaining parts may be COTS products, or products conducted by subcontractors.

3 For detailed description of problem description templates, please read Kajko-Mattsson (2001a).

REFERENCES

American National Standard Institute/IEEE (ANSI/IEEE STD-982.2). (1988). *Guide for the Use of IEEE Standard Dictionary of Measures to Produce Reliable Software.* Los Alamitos, CA: Computer Society Press.

American National Standard Institute/IEEE (ANSI/IEEE STD-610.12). (1990). *Standard Glossary of Software Engineering Terminology.* Los Alamitos, CA: Computer Society Press.

American National Standard Institute/IEEE (ANSI/IEEE STD 1219). (1998). *Standard for Software Maintenance.* Los Alamitos, CA: Computer Society Press.

Annick, P. G. (1993). Integrating natural language processing and information retrieval in a troubleshooting help desk. *IEEE Expert, 8*(6), 9-17.

Arthur L. J. (1988). *Software Evolution: The Software Maintenance Challenge.* New York: John Wiley & Sons.

Boehm, B. W. (1973). *The High Cost of Software.* Paper presented at the Symposium on High Cost Software.

Bouman, J., Trienekens, J., & van der Zwan, M. (1999). *Specification of Service Level Agreements, Clarifying Concepts on the Basis of Practical Research.* Paper presented at the conference of Software Technology and Engineering Practice.

Briand, L. C., Basili, V. R., Yong-Mi, K., & Squier, D. R. (1994). *A Change Analysis Process to Characterise Software Maintenance Projects.* Paper presented at the IEEE International Conference on Software Maintenance.

Cashman, P. M., & Holt, A. W. (1980). A communication-oriented approach to structuring the software maintenance environment. *ACM SIGSOFT, 5*(1), 4-17.

Chapin, N. (2001). Types of software evolution and software maintenance. *Journal of Software Maintenance and Evolution, 13*(1), 1-30.

DeRoze, B., & Nyman, T. (1978). The software life cycle—A management and technological challenge in the Department of Defense. *IEEE Transactions On Software Engineering, 4*(4), 309-318.

Florac, W. A. (1992). *Software Quality Measurement: A Framework for Counting Problems and Defects* (Report no. CMU/SEI-92-TR-22). Software Engineering Institute, Carnegie-Mellon University, Pittsburgh, PA.

Foster, J. R., Jolly, A. E. P., & Norris, M. T. (1989). An overview of software maintenance. *British Telecom Technology Journal, 7*(4), 37-46.

Glass, R. L., & Noiseux, R. A. (1981). *Software Maintenance Guidebook.* New York: Prentice-Hall.

Hart, S. (1993). *Technical Support: A Valuable Voice.* Paper presented at the International Professional Communication Conference.

Hazeyama, A., & Hanawa, M. (1999). *A Problem Report Management System for Software Maintenance.* Paper presented at the IEEE International Conference on Systems, Man and Cybernetics.

Jones, C. (1994). *Assessment and Control of Software Risks.* Englewood Cliffs, NJ: Yourdon Press.

Kajko-Mattsson, M. (1999a). *Software Problem Resolution at ABB Automation Products* (Report no. 99-003). Department of Computer and Systems Sciences (DSV), Stockholm University and Royal Institute of Technology, Sweden, Kista.

Kajko-Mattsson, M. (1999b). *Software Problem Resolution at ABB Robotics AB*

(Report no. 99-004). Department of Computer and Systems Sciences (DSV), Stockholm University and Royal Institute of Technology, Sweden, Kista.

Kajko-Mattsson, M. (1999c). *Maintenance at ABB (I): Software Problem Administration Process (The State of Practice)*. Paper presented at the IEEE International Conference on Software Maintenance.

Kajko-Mattsson, M. (1999d). *Maintenance at ABB (II): Change Execution Processes (The State of Practice)*. Paper presented at the IEEE International Conference on Software Maintenance.

Kajko-Mattsson, M. (2000). *Preventive Maintenance! Do We Know What It Is?* Paper presented at the IEEE International Conference on Software Maintenance.

Kajko-Mattsson, M. (2001a). Corrective Maintenance Maturity Model: Problem Management. *Dissertation* (ISBN Nr 91-7265-311-6, ISSN 1101-8526, ISRN SU-KTH/DSV/R - 01/15). Department of Computer and Systems Sciences (DSV), Stockholm University and Royal Institute of Technology, Sweden, Kista.

Kajko-Mattsson, M. (2001b). *CM^3: Upfront Maintenance.* Paper presented at the Conference on Software Engineering and Knowledge Engineering, Knowledge Systems Institute.

Kajko-Mattsson, M. (2001c). *Towards a Business Maintenance Model.* Paper presented at the IEEE International Conference on Software Maintenance.

Kajko-Mattsson, M. (2001d). *Motivating the Corrective Maintenance Maturity Model (CM^3).* Paper presented at the Seventh IEEE International Conference on Engineering of Complex Computer Systems.

Kajko-Mattsson, M. (2001e). *A Glimpse into CM^3: Problem Management.* Paper presented at the Conference on Software Advanced Information Systems Engineering, Springer-Verlag.

Kajko-Mattsson, M. (2001f). *Can We Anything Learn from Preventive Hardware Maintenance?* Paper presented at the Seventh IEEE International Conference on Engineering of Complex Computer Systems.

Kajko-Mattsson, M. et al. (2001). *The State of Documentation Practice.* Paper presented at the IEEE International Conference on Software Maintenance.

Kajko-Mattsson, M., Forssander, S., & Andersson, G., (2000a). Software Problem Reporting and Resolution Process at ABB Robotics AB: State of Practice. *Journal of Software Maintenance and Evolution, Research and Practice, 12*(5), 255-285.

Kajko-Mattsson, M., Forssander, S., Andersson, G., & Olsson, U. (2002). Developing CM^3: Maintainer's Education and Training at ABB. *Journal of Computer Science Education,* 3.

Kajko-Mattsson, M., Forssander, S., & Olsson, U. (2001). *Corrective Mainte-*

nance Maturity Model (CM³): Maintainer's Education and Training. Paper presented at the IEEE International Conference on Software Engineering.

Kajko-Mattsson, M., Grimlund Glassbrook, A., & Nordin, M. (2001i). *Evaluating the Predelivery Phase of ISO/IEC 14764 in the Swedish Context.* Paper presented at the IEEE International Conference on Software Maintenance.

Kaner, C., Falk, J., & Nguyen, H. Q. (1999). *Testing Computer Software*. New York: John Wiley & Sons.

Lee, I., McRee, R., & Barlett, W. (1996). *On-Line Recovery for Rediscovered Software Problems*. Paper presented at the IEEE International Computer and Dependability Symposium.

Martin, J., & McClure, C. (1983). *Software Maintenance, The Problem and Its Solutions*. Englewood Cliffs, NJ: Prentice-Hall.

Mills, H. D. (1976). Software Development. *IEEE Transactions on Software Engineering, 2*(4), 265-273.

Myers, G. J. (1976). *Software Reliability Principles and Practices*. New York, John Wiley & Sons.

Niessink, F. (2000). Perspective on Improving Software Maintenance. *Dissertation* (No. 2000-1). SICS, Vrije University, Holland.

Nihei, K., Tomizawa, N., Shibata, A., & Shimazu, H. (1998). *Expertguide for Help Desks–An Intelligent Information Retrieval System for WWW Pages.* Paper presented at the 9th International Workshop on Database and Expert Systems Applications.

Parihk, G. (1986). *Handbook of Software Maintenance A Treasury of Technical and Managerial Tips, Techniques, Guidelines, Ideas, Sources, and Case Studies for Efficient, Effective, and Economical Software Maintenance.* New York: John Wiley & Sons.

Pigoski, T. M. (1997). *Practical Software Maintenance*. New York: John Wiley & Sons.

Powersoft. (1997). *Power Designer Data Architect, User's Guide.* Sybase: Enterprise.

Schneidewind, N. (1987). The state of software maintenance. *IEEE Transactions on Software Engineering, 13*(3), 303-310.

Swanson, B. E. (1999). IS maintainability: Should it reduce the maintenance effort? *ACM SIGCPR, 4,* 164-173.

Wedde, K. J., Stalhane, T., & Nordbo, I. (1995). *A Case Study of a Maintenance Support System*. Paper presented at the IEEE International Conference on Software Maintenance.

Zuther, M. (1998). *Modular Based IT-Solutions for Efficient World-wide Service Processes.* Paper presented at the 24th Annual Conference of the IEEE Industrial Electronics Society.

APPENDIX
MODELING CONSTRUCTS

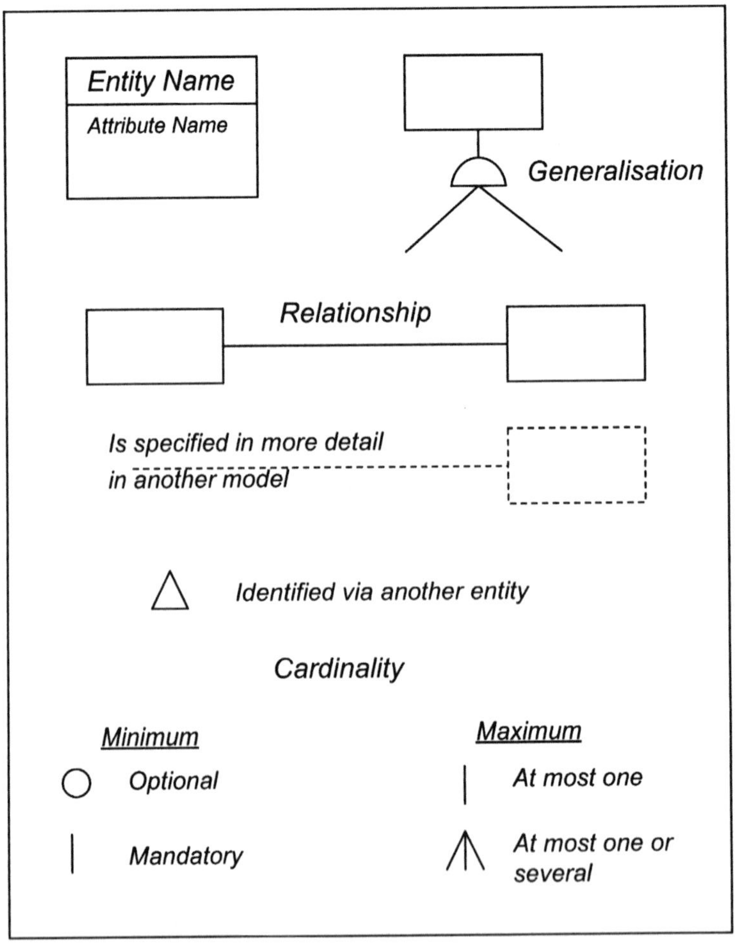

Entity Name
Attribute Name

Generalisation

Relationship

Is specified in more detail
in another model

Identified via another entity

Cardinality

Minimum

Optional

Mandatory

Maximum

At most one

At most one or several

Chapter III

The Impact of eXtreme Programming on Maintenance

Fabrizio Fioravanti
Finsystem s.r.l., Italy

One of the emerging techniques for managing software project is eXtreme Programming (XP), which surely is changing the way in which we develop and manage software. XP is based on four values and 12 rules that explain how to develop software in an XP-compliant mode. In this chapter, all values and rules are addressed and the XP lifecycle of a project is introduced in order to guide the reader in discovering how XP can aid maintenance by keeping maintenance costs constant over time. Before drawing conclusions, the direct impact on the maintenance process due to the adoption of XP values and rules is also exploited at the end of the chapter.

EXTREME PROGRAMMING: A GENTLE INTRODUCTION

One of the emerging techniques for managing software project is eXtreme Programming (Beck, 1999 and 2000). XP surely changes the way in which we develop and manage software, but in this chapter we will explore also how it can change the way we maintain software. The most interesting feature of eXtreme Programming is the fact it is "human oriented" (Brooks, 1995). XP considers the human-factor as the main component for steering a project towards a success story. It is important to evidence that XP is guided by programming rules, and its more interesting aspects deal with the values that are the real guides for the design, development and management processes.

The values on which XP is based are:

- **Communication**: it is the first value, since XP's main objective is to keep the communication channel always open and to maintain a correct information flow. All the programming rules I will describe later cannot work without communication among all the people involved in the project: customers, management, and developers. One of the roles of management is to keep communication always up (Blair, 1995).
- **Simplicity**: the first and only question that you have to answer, when encountering a problem is: "What is the simplest thing that can work?" As an XP team member, you are a gambler, and the bet is: "It is better to keep it simple today and to pay a little bit tomorrow for changing it, instead of creating a very complicated module today that will be never used in the future." In real projects with real customers, requirements change often and deeply; it is crazy to make a complex design today that have to be completely rewritten in one month. If you keep it simple initially, when the customer changes his mind or when the market evolution requires a change, you will be able to modify software at a lower cost.
- **Feedback**: continuous feedback about the project status is an added value that can drastically reduce the price that you must pay when changes have to be made. Feedback is communication among management, customers and developers, that is, direct feedback; feedback is also tracking and measuring, so that you can maintain control of the project with indirect measures on the code, on its complexity, and on the estimated time to make changes. Feedback is not only communication but also deals with problems of software measurement and then with metrics.
- **Courage**: courage is related to people, but it is always related to the way in which you develop your project. XP is like an optimization algorithm, such as simulated annealing in which you slightly modify your project in order to reach a optimal point. But when the optimum you reach is only local, you need the courage to make a strong change in order to find the global optimum. XP with its rules encourages you to re-factor when your project is stalled, but in order to re-factor in deep you need courage.

These values are the bases for the 12 XP rules. The following is a short description of these rules.

Planning Game

During software development, commercial needs and technical considerations have to steer the project in equivalent ways. For these reasons, the planning of an XP project is decomposed in several steps. Starting from a general metaphor, for each functionality a *story* is written; a story summarizes what the function has to do and what results have to be obtained from it. Each story is then divided into a variable number of tasks that are in example classes or part of them. Stories are then sorted by relevance, where relevance is related to the benefit to the user and to the added value obtained by the implementation of the story for the whole project. Within each story, tasks are then sorted by relevance, and the more important are selected by the manager in order to choose the subfunctionality to be implemented first. These are the ingredients for the recipe that the planning game is.

Planning has to be treated as a game in which each person plays a role and has a specific assignment to be carried out. Managers have to make decisions about priorities, production deadlines, and release dates, while developer have to: estimate the time for completing assigned tasks, choose the functionality that has to be implemented first among those selected by the customer for being implemented in the current release, and organize the work and the workspace in order to live better.

Small Releases

Each software release should be given out as soon as possible with the functionality of greater value correctly working. A single release is comprised of several subreleases that implement single functionalities that in XP are called tasks. More tasks collected together to form a more complex and a self-comprehending functionality determines a story that can be implemented during a release. Note that adopting this operating way, it is not possible to implement only a half of a functionality, but only to shorten the release time.

Metaphor

Each XP project is summarized as a metaphor or a collection of metaphors that are shared among customers, management, and developers. The metaphor aids the customer to describe in natural language what he wants, and if refined by technical persons it can be a good starting point for defining requirements. The metaphor that can be assumed, for example, for a XML editor is that of a word processor (that means cut, copy and paste operations plus file loading/saving and the capability of writing something) that is able to write only tags with text inside. The vocabulary of the editor is the schema or DTD set of tags, the spell checker is the XML well-formedness and the grammatical check is the validation against DTD or schema.

Tests

Programmers write unit tests in order to be confident that the program will cover all the needed functionality; customers write functional tests in order to verify high-level functionality. You don't have to necessarily write one unit test for each method, but you have to write a test for each part of the system that could possibly break now or in the future. Remember to also address the problem of monkey test in which the system has to deal with crazy data in input.

Simple Design

Each part of the system has to justify its presence by means of its used functionality. In general, the simpler project in each time instant (i) makes all the tests run, (ii) has no code or class duplication, (iii) is quite self-documenting by adopting explicatory names and commenting on all the functionality, and (iv) has the least possible number of classes and methods to correctly implement the functionality.

Refactoring

Before adding a new functionality, verify if the system can be simplified in order to make less work to add it. Before test phases, programmers ask themselves if it is possible to simplify the old system in order to make it smarter. These techniques are at the basis of continuous refactoring. Keep in mind that refactoring is carried out to simplify the system and not to implement new functions that are not useful.

Pair Programming

All the code must be written by two programmers in front of the same computer (Spinellis, 2001; Williams, Kessler, Cunningham, & Jeffries, 2000). This can be a good way to instruct new team members, reducing skill shortage and, in all cases, is a good practice to increase productivity and reduce error probability, since one person is continuously stimulated by the other to produce quality code and to correct the errors that the other inserts in the program. Also a not-so-skilled programmer can be a valid aid in pair programming since he can ask questions of the more skilled person that can produce a better testing suite. Feedback helps the less skilled to improve his system knowledge, and the right question at the right moment can be a contribution for the more skilled to find a solution that can simplify the code.

Continuous Integration

The code is tested and integrated several times a week and in some cases a few times a day. This process improves the probability of detecting hidden faults in other parts of the system due to the modification planned. The integration ends only when all the previous and the new tests work correctly.

Collective Ownership

Whoever sees the possibility to simplify a part of the code has to do it. With individual code property, only the person that has developed a part is able to modify it, and then the knowledge of the system decreases rapidly. Consider the possibility that the code-owner accepts a job with another company; in that case ... you are really in trouble without the adoption of collective ownership.

On-site Customer

The customer has to be present in the workgroup and has to give feedback to developers. He also has to write functional tests in order to verify that the system works correctly and has to select the functions that add greater value to the system in order to choose which functionalities should be present in the next release of the software.

40-Hour Weeks

You cannot work for 60 hours a week for a long time. You'll become tired and stressed and your code will have a lower quality. In addition, your capability to interact with other people will decrease (Brooks, 1995).

Coding Standards

The adoption of a coding standard, so that all programmers will have the same way to write code, minimizes the shock due to seeing a code not formatted to your personal standards. In particular, comments, especially those for automated documentation generation, have to be standardized; parentheses and indentation should be uniform among all the code in order to facilitate comprehension and collective ownership.

In the following, these rules will be discussed from the point of view of maintenance, evidencing that XP can help achieve a constant maintenance cost over time.

The XP rules and values have a strong impact on the life cycle of the project, and the life cycle itself drives the maintenance activities, as will be shown in the following. For that reason, it is necessary to introduce the XP life cycle and to consider its impact on maintenance.

LIFE CYCLE OF AN XP PROJECT

Evolutionary programming paradigms, such as spiral-based ones, make an assumption about activities to be performed during the development of a task. Classically, the phases that are considered during task development are the following: Requirement, Analysis, Design, Code, Test, and Integration.

Considering that each task is partially superimposed on the previous one and the next, a graph as shown reported in Figure 1 can be drawn for effort assignment.

The life cycle of an XP task is slightly different because of the approach to program development using XP rules. The phases of task development in XP are:

- Requirements and analysis: these tasks are usually performed together since the customer is on-site, and the feedback about functionalities to be implemented is faster and more precise. This reduces the effort and the time for these operation compared to a more classical approach.
- Design: in an XP project, we never project for tomorrow, but we only address the problems related to the task selected by the on-site customer. This reduces the time-window, so that the effort for this phase can be better employed for other task phases.
- Refactoring: continuous refactoring is one of the rules of XP and a refactoring phase has to be performed before any modification is introduced in the system. During this phase, the system is analyzed in order to determine if the possibility of system simplification exists. This task can take more time than the design phase since all the subsystems related with the task under development have to be revised and eventually improved and/or simplified. Since small releases are a must in XP, the refactoring activities are very close to each other, approximating a continuous refactoring.
- Write-Tests: during this phase, the classes/functions of the systems needed to implement the new functionalities are realized in the form of a skeleton, and the code inserted is only for the purpose of satisfying, as with dumb code, the unit tests we have to write in this phase. Unit tests must cover all the aspects that have the probability of not working.

Figure1. Effort during evolutionary life cycle (i.e., spiral).

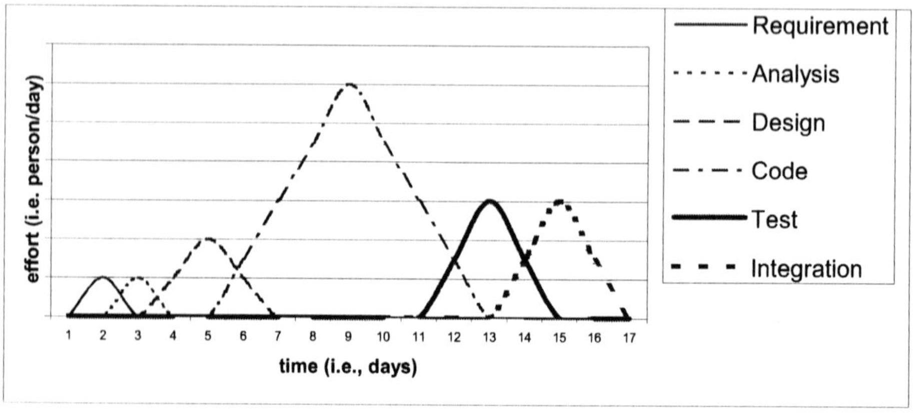

- Test-Integration: in this phase, the integration of all the classes developed in the previous phase are integrated in the system and it is determined if all the unit tests that were developed in the past continue to work. The effectiveness of this phase deals with the opportunity to verify that the design of the new functionality will not have an adverse impact on other system operations that were stable in the past.
- Code: in this phase, the code for the new functionalities is inserted into the skeleton provided by the write-test phase. Several phases of this kind have to be carried out before the final release of the task. Each coding phase has to be followed by a test and integration step.
- Integration and test: during this phase, it is determined if all the previous unit-tests and the functional tests work correctly. If all are OK, then the code is approved and is considered stable.

Following the guidelines of the XP life cycle, the graph in Figure 2 can be drawn. The two graphs (Figure 1 and Figure 2) have been depicted keeping the total effort and the duration of the task constant.

The trend of cumulative effort vs. time unit for the two approaches are depicted together in Figure 3. Note that the areas under the two graphs are the same. This assumption is made in order to show that the total effort for developing a task can be considered constant independently of the approach chosen. This assumption is relevant considering the maintenance effort trend that will be examined in the next section, which will show that XP adoption can result in a constant maintenance effort during the life cycle of the task instead of an exponential trend. The maintenance effort can be also estimated to be lower if an XP approach to software development is considered.

Figure 2. Effort during XP life cycle.

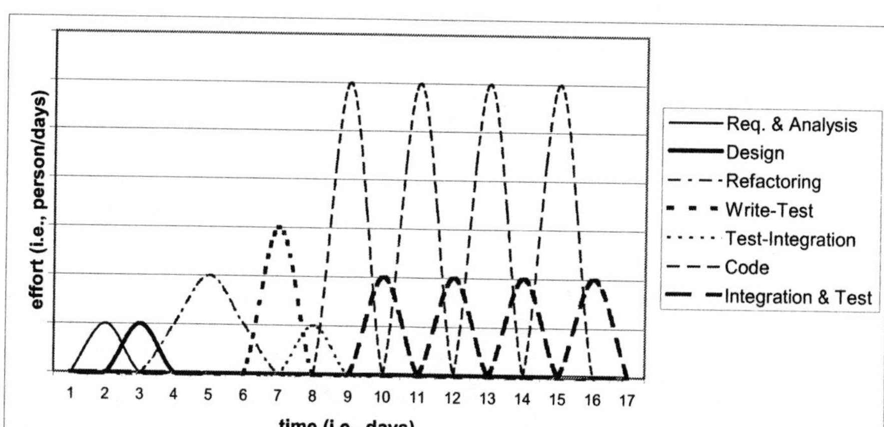

THE PROMISE OF XP: A CONSTANT MAINTENANCE COST

The promise of XP, and at the same time the starting point of the XP approach, is that the maintenance cost during the life cycle of the project is quite constant. In all the software engineering books, you find that costs of changes raise in an exponential way during time, as depicted in Figure 4 (Bohem, 1981; DeMarco, 1982; Gilb, 1988).

This hypothesis impacts in the way we project and code, since if the maintenance cost really raises as an exponential, we have to project today for reusing tomorrow; we have to implement and study as many functionalities as possible in the early phases of the project in order to have a lower number of modifications in the future, and so on (Jacobson, 1992).

But imagine for a while that the maintenance cost curve has an horizontal asymptote, as in Figure 5. What about your programming practices? How do you change the way in which you project, develop and manage?

If this promise is true then you'll never do today what *maybe* will be useful tomorrow, since with a very low overhead you can add the functionality "if and only if" it is required in the future. You'll never make a decision now if you can postpone it until your customer really asks for that decision and when the time has reduced the risk of making the wrong decision with respect to an early phase of software development, and so on. The approach in which you develop drastically changes if you have the curve of Figure 5 in mind when you think of maintenance costs.

It is important to verify, almost from a theoretical point of view, if the XP promise about maintenance costs can be considered realistic or not.

Figure 3: Cumulative effort per time unit.

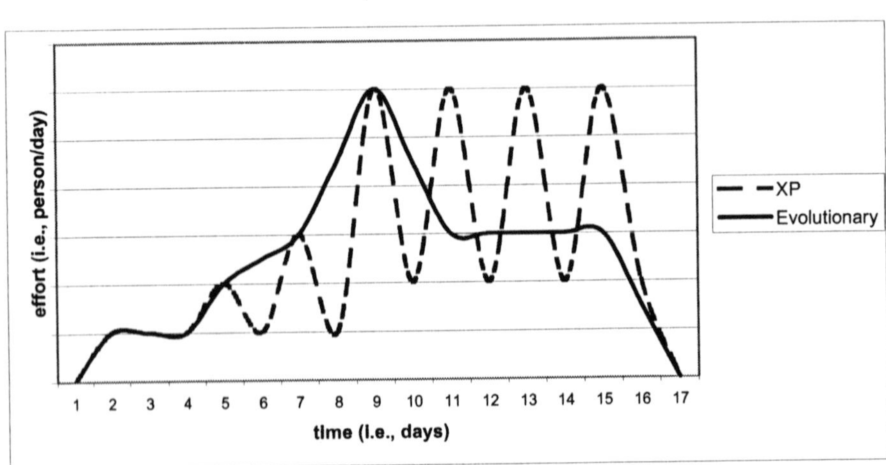

In Figures 6 and 7, the testing points for measuring the needed maintenance effort for each approach are shown by arrows showing the testing points. For each point the maintenance effort is calculated supposing that an error will be detected at that time. With respect to the time scale reported in the figures, we assume the following notation, ERR@t means that we discover an error at time *t*. A detailed description of the error's impact on the previous phases is reported during the discussion of Tables 1 and 2.

Considering Figure 6, the errors detected and the phases they impact for a certain percentage of the phase effort are evidenced in Table 1.

An explanation of the table is needed in order for the reader to better understand the reported numbers:

- ERR@3: An error detected at middle of the analysis phase impacts partially (50%) a review of the requirement, and also the already performed analysis (and then a 25% impact has been supposed).
- ERR@5: As for the analysis phase, we can suppose that an error detected during the design impacts the maintenance effort in a similar manner, and then for 50% of requirements and analysis, and for 25% of design.

Figure 4: Maintenance cost increases exponentially during time in a classical software engineering approach.

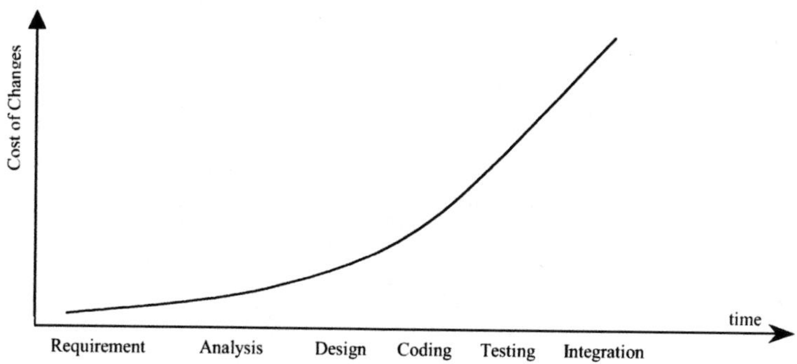

Figure5: Maintenance cost in XP approach.

- ERR@9: The same consideration already performed before can be redone also for errors detected during this phase.
- ERR@13: In this case, the error is detected during a test phase and then it can impact all the previous phases, but in order to verify that the correction has been performed correctly, all the testing work has to be redone. For this reason, we have an impact of 50% on the test phase (that is, the effort spent for testing at instant 13).

Figure 6: Cumulative effort for evolutionary development with error detection point highlighted.

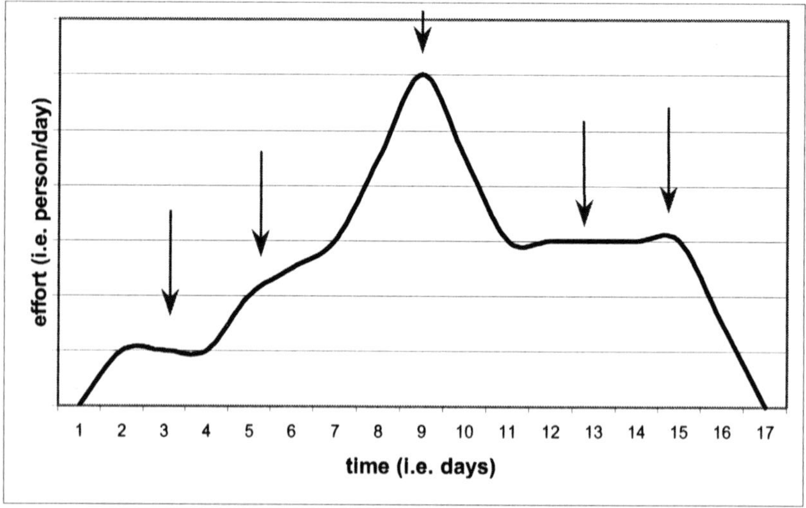

Figure 7: Cumulative effort for XP with error detection point highlighted.

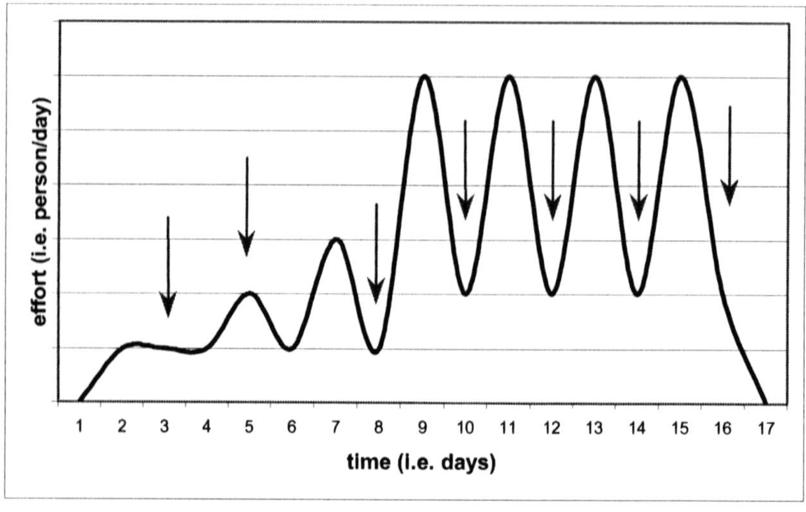

- ERR@15: The same consideration as for testing can be rewritten paying attention to the fact that all the tests and the integration work already performed have to be repeated.

Considering these values, the trend of maintenance efforts with respect to time is drawn in Figure 8. The trend depicted confirms the classical shape of the maintenance effort during life cycle reported in Figure 4.

Let me repeat the same considerations related now to Figure 7 considering the marked error points. Before presenting a similar table with respect to that for the evolutionary life cycle, let me introduce the main innovative concepts hidden behind the XP life cycle. In Figure 6 five integration and testing points are evidenced; this means that if a test phase has been successfully passed, then the preceding phases

Table 1: Maintenance effort with respect to the already spent effort expressed as a function of the time instant in which an error is identified for classical evolutionary life cycle.

Error	Requirement	Analysis	Design	Code	Test	Integration
ERR@3	50%	25%	0%	0%	0%	0%
ERR@5	50%	50%	25%	0%	0%	0%
ERR@9	50%	50%	50%	25%	0%	0%
ERR@13	50%	50%	50%	50%	50%	0%
ERR@15	50%	50%	50%	50%	100%	50%

Figure 8: Maintenance costs during the evolutionary system life cycle.

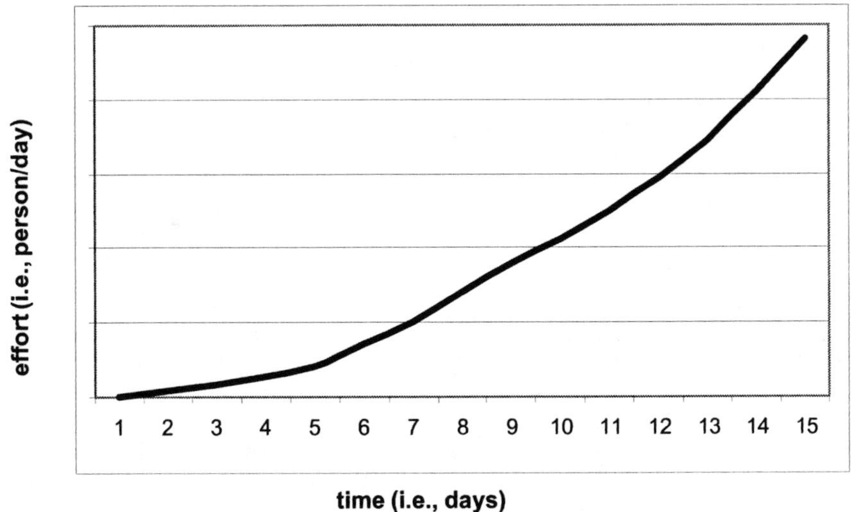

can be considered stable. This assumption is true since at each test/integration phase all the unit and functional tests for the system are executed, and then the written code and the analysis-design phase can be considered well done.

By XP adoption, errors can be discovered during design and refactoring phases and during test-integration phases, and therefore, considering as stable the phase before a successful test, Table 2 can be drawn.

Also in this case, an explanation of the table is needed for the reader to better understand the reported numbers:

- ERR@3: An error detected at the middle of the design phase partially impacts (50%) on a review of the requirement and on the analysis, and also on the already performed design phase (and then a 25% impact has been supposed).
- ERR@5: As for the design phase, we can suppose that an error detected during refactoring impacts in a similar manner on the maintenance effort, and then for 50% of requirements, analysis and design phases, and for the 25% of refactoring, in the mean.
- ERR@8: The same consideration already performed before can also be redone for errors detected during this phase, considering also that all the already performed work on testing should be repeated in order to be sure of delivering a stable version of the re-factored system.
- ERR@10, ERR@12, ERR@14, ERR@16: In this case (ERR@10), the error is detected during a test phase. Since the re-factored code can be considered stable because of the already performed tests, the error has probably been inserted in the last phase of coding. This means that the written code since the last successful test phase should be partially (50%) revised, and the test phase has to be completely redone. Similar considerations can be repeated for testing points at times 12, 14, and 16.

Table 2: Maintenance effort with respect to the already spent effort, expressed as a function of the time instant in which an error is identified for XP life cycle. R&A means Requirement and Analysis; D means Design; R means Re-Factoring; WT means Write Tests; TI means Test Integration; C_n is the nth Coding Subphase; I&T_n the nth Integration and Test phase.

Error	R&A	D	R	WT	TI	C_1	I&T_1	C_2	I&T_2	C_3	I&T_3	C_4	I&T_4
ERR@3	50%	25%	0	0	0	0	0	0	0	0	0	0	0
ERR@5	50%	50%	25%	0	0	0	0	0	0	0	0	0	0
ERR@8	50%	50%	50%	50%	50%	0	0	0	0	0	0	0	0
ERR@10	0	0	0	0	0	50%	100%	0	0	0	0	0	0
ERR@12	0	0	0	0	0	0	0	50%	100%	0	0	0	0
ERR@14	0	0	0	0	0	0	0	0	0	50%	100%	0	0
ERR@16	0	0	0	0	0	0	0	0	0	0	0	50%	100%

The graphical representation of Table 2, on the basis of the consideration previously made, is reported in Figure 9.

From a theoretical point of view, it can be evidenced that a constant maintenance costs can be achieved by adopting XP, and that maintenance costs can be kept under control in a better way, if we double the effort of Table 2. The real question is how to obtain such a trend in a real project? For sure, you have to satisfy some general principles, such as:

- Keep the project simple and clean. Do not add anything that "maybe" will be used in the future; add only the functionality that you, and especially the customers, need. All the parts of the project have to be useful, otherwise they have to be removed or simplified.
- Use automatic unit and functional test to easily verify if a change impacts in same hidden way on other parts of your project. It is also the way to implement continuous regression testing.
- All the team members have to share knowledge about the project so that each person knows how to modify the code without strong overhead costs for re-analysis of the project and of the already written code.
- Integrate your part as soon as possible and do it only when all the new and previous tests work correctly.

The application of the rules of XP are very helpful in obtaining the previously cited points and, therefore, it is a deep push in the direction of constant maintenance costs, as evidenced in the next section where the impact of XP rules on maintenance will be discussed.

Figure 9: Maintenance costs in XP life cycle (the y-axis scale is the same with respect to that of Figure 8).

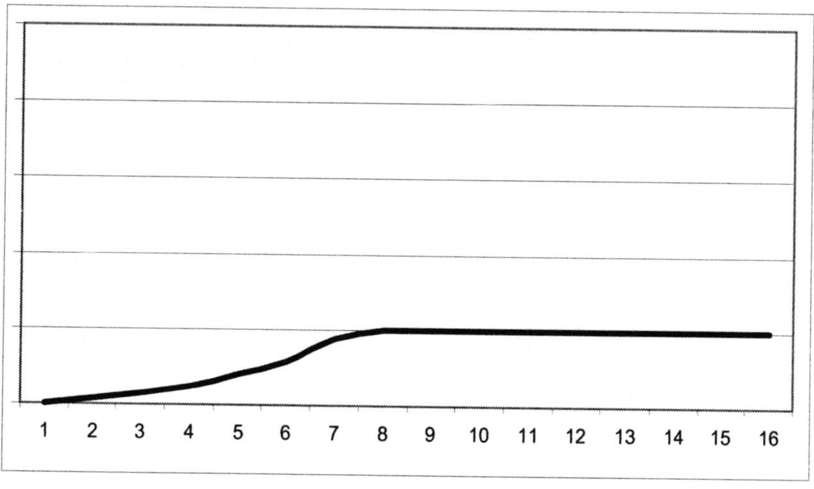

The innovative concept is that the XP adoption in a project constantly keeps the project in the maintenance status. The cost of changes when you follow the classical rules for project development is strongly influenced by the fact that you postpone the integration and production phases. If you are in production since the first phases of the project, your mind is always involved with maintenance problems, since maintenance is the normal status of the project itself, while developing is only a transient and as short as possible period of your project lifetime.

XP IMPACT ON MAINTENANCE

In this section, the impact of the XP rules and practices on maintenance will be evidenced. The rules and the values on which rules are based will be examined in order to highlight the added value that each practice can give to the maintenance process.

Communication

Good communication means that all team members know exactly what is happening on the code; that is, each person can modify each part of the system. If he/she is not able to do so with his partner, he/she will look for another partner for the specific task. The impact of this value on maintenance is evident—the reduction of time-to-production for each team member. Communication value strongly helps to push down the modification cost curve, since no proprietary part of code exists.

The communication among team and customers also helps to identify errors and misunderstandings about the project, as soon as possible minimizing the time spent for useless or unnecessary parts of code.

Simplicity and Simple Design

A simple project with all the classes having a specific function and scope aids the identification of the parts to be modified in a generic maintenance activity, decreasing the related maintenance costs. Changes due to modified market needs or customer requirements can be addressed in a faster way if the project and code are kept simple and clean

Feedback

Feedback between customer and those who understand the software enhances the probability of identify the greatest part of the problems. The direct contact with customers, who also write functional tests, is a strong push in the direction of a continuous preventive maintenance activity that allows a reduction in global costs of the whole maintenance process.

Courage

Courage, in the sense of XP, can improve the simplicity of a system and therefore reduce the maintenance cost and improve the effectiveness of maintenance activities.

Courage, especially in the refactoring phase, allows a milestone reducing, as previously shown, the future maintenance costs.

Planning Game

Well, a smart management is always an aid during all the phases of software development, and therefore during maintenance. Consider also the fact that the functionality with the highest added-value is developed first and then tested in many different operating conditions, allowing the delivery of a more stable code for at least the main functionalities. The planning game (small tasks to be implemented and tested in sequence) impacts strongly on the life cycle and then on maintenance costs.

Small Releases

Adding only a few functionalities at a time and verifying each release with the customer, reduces the risk of needed corrective maintenance to be carried out. Smaller releases more closer in time can improve the problem solving in the sense that if a fault is present, it will be eliminated more quickly reducing the time to market of a stable software version.

Metaphor

A general guideline that is clear for all the persons involved in the software development and test activities helps to focus the project aims and scope. Metaphor has no direct consequences on maintenance, but the idea of a metaphor to be implemented as a story and as tasks change the life cycle of a project in significant ways.

Tests

It is obvious that testing reduces the probability of future appearances of undiscovered errors. From the point of view of maintenance, automated unit and function tests allow the identification of the problems inserted by the changes performed on code. Regression testing is universally recognized as an activity to reduce future maintenance cost during corrective maintenance, and regression testing is the usual practice in XP.

Refactoring

Maintenance for the sake of maintenance, also called perfective maintenance, is not universally recognized as a good practice. When it is used to enforce project simplicity and reduce the probability of future corrective maintenance activities it has to be considered a good rule to be followed. XP continuously refactors with this aim.

Pair Programming

It is evident that four eyes see better than two, but the greater advantage of pair programming is the sharing of knowledge among the team members. The reduction of maintenance costs is obtained by diminishing the time needed for programmers to understand a part of code written by other people, since all the advantages of collective code ownership are a direct consequence of this particular technique for developing.

Continuous Integration

The addition of small parts of code to a stable version reduces the probability of inserting errors since the code can be integrated only when all the previous tests continue to work correctly. This modus operandi reduces the probability of inserting indirect errors in other parts of the system and then curtails the costs related to corrective maintenance, since these activities are performed directly during integration, in case of failure of the regression testing.

Collective Ownership

Collective ownership of code obviously facilitates the sharing of knowledge, but the main advantage with respect to maintenance is due to the need to adopt a revision control system for storing the code. In the case of error, the project can be reverted to the last stable release of the problematic module, giving to the customer the impression that maintenance activities can be performed without a great impact on the installed systems and with minimal reduction in functionalities.

On-site Customer

The presence of the customer on-site improves the maintenance factors related to functional tests and feedback. The customer select the more relevant stories for the project. These stories are realized first, and then they are tested and re-factored many times. These activities give more stability and simplicity to the functions that are more relevant for the customer, increasing customer satisfaction and reducing the probability of project rejection.

40-Hour Weeks

Extra work is not good for code quality, nor for maintenance activities. This can be trivial, but it is true.

Coding Standards

A better comprehension of the source code simplifies the change process and thus reduces the time needed for maintenance activities. A standardized way to write comments allows also automatic documentation that is improved each time a comment line is added.

CONCLUSION

From all the examined features it can be easily determined that maintenance costs can be kept under control if all the previous rules are applied correctly and with the right aim. Constant maintenance costs during the software life cycle is no more a chimera; it is not so easy to be achieved, but with adoption of the right programming practices, a good approximation can be reached. XP has no "silver bullets" for reducing maintenance costs; however, each rule enforces all the others, creating an environment in which maintenance is the normal status of a project, and then all people involved work to reduce the costs related to the main activity as much as possible.

The major disadvantage of XP is its applicability for industrial projects. From my experience and point of view, a large effort is needed to convince programmers to work in couples, exploiting pair programming, and to change the way in which they program, i.e., writing tests before implementing code and standardizing the way in which they write code comments and documentation. One of the rules of thumb for the success of XP techniques is the cooperative work among people, and from my experience, cooperative work can be successfully achieved in a group not larger than 10 programmers. Then, you can control the maintenance costs with the adoption of XP, but you must be very careful to deploy XP techniques in a group that is ready to apply its rules.

REFERENCES

Beck, K. (1999). Embracing change with extreme programming. *IEEE Computer.*

Beck, K. (2000). *Extreme Programming Explained: Embrace Change.* Boston, MA: Addison Wesley Longman.

Blair, G. M. (1995). Starting to manage: The essential skills. *IEEE Engineers Guide to Business Series.*

Bohem, B. (1981). *Software Engineering Economics.* Englewood Cliffs, NJ: Prentice-Hall.

Brooks, F. (1995). *The Mythical Man-Month.* Reading, MA: Addison-Wesley.

DeMarco, T. (1982). *Controlling Software Projects.* Englewood Cliffs: Yourdon Press.

Gilb, T. (1988). *Principle of Software Engineering Management*. Reading, MA: Addison-Wesley.

Jacobson, I. (1992). *Object-Oriented Software Engineering: A Case Driven Approach*. Reading, MA: Addison-Wesley.

Spinellis, D. (2001). Fear of coding, and how to reduce it. *IEEE Computer*.

Williams, L., Kessler, R.R, Cunningham, W., & Jeffries, R. (2000). Strengthening the case for pair programming. *IEEE Software*.

Chapter IV

Patterns in Software Maintenance: Learning from Experience

Perdita Stevens
University of Edinburgh, Scotland

In software design, patterns—that is, structured, named descriptions of good solutions to common problems in context—have become a popular way of recording and transferring problem-solving expertise. The aim of this chapter is to describe how patterns can help in the field of software maintenance.

There are two main uses of patterns in software maintenance. The first is to consider the introduction of a design pattern to an existing design. The second approach is to consider patterns for the maintenance process itself. For example, reengineering patterns describe solutions to common problems in the reengineering of a system, considering issues such as how to split the necessary changes into risk-minimizing steps. We discuss the advantages, pitfalls and practicalities of using patterns in both of these ways.

INTRODUCTION

Software maintenance is widely seen as the "poor relation" of the software engineering world. The skills that it requires are less widely recognized than those needed by more glamorous areas such as design or even testing. Few books on software maintenance are available, and those that exist are seldom seen on bookshops' shelves. Software maintenance is gradually being recognized as an important area to cover in software engineering degree courses, but even now it is often covered in just a few hours of teaching. The overall effect is that maintainers of software generally acquire their skills almost entirely on the job.

It is difficult to come up with a convincing explanation of why this should be so, and fortunately that is not the focus of this chapter. Rather, we focus here on how to move forward from where we are; taking the absence of a large, commonly-agreed upon body of software maintenance knowledge as given, how can an organization nevertheless ensure that it makes best use of the knowledge and experience available to it? In particular, we explore how people and organizations undertaking software maintenance can use *patterns* as a technique to help them learn from experience—either their own experience, or that of others.

What do we mean by a "pattern" in this context? We will go into more detail in the next section, but for now it suffices to say that a pattern is a named, structured description of a good way to solve a common problem. It's a codified version of somebody's experience, or preferably of several different people's experience, with all the benefits and drawbacks which that implies, which we will discuss later. It typically takes up a few paragraphs to a few pages of text. An organization that is "using patterns" may be consulting patterns that have been written elsewhere, or it may be using the writing of patterns as a way of recording its own lessons learned, or (perhaps best) both. It is important to realize that in either case, the use of patterns will not constitute a complete methodology; you would not expect to welcome new recruits by giving them the company's pattern library and nothing else. Patterns are most useful when they are used alongside a company's existing methodology and process and integrated with it, and their use can be integrated with just about any process that a maintenance organization might use. Later in this chapter, we will return to the question of how to integrate the use of patterns with your maintenance process.

It is the flexibility and scalability of pattern use—the fact that a little or a lot of expertise may be recorded as patterns, and the fact that pattern use can supplement, rather than replace, what already works—that makes patterns interesting in the area of software maintenance. Whatever your organization's software maintenance maturity, this chapter hopes to convince you that it is worthwhile to consider using patterns in taking the next step. Thus the objectives of this chapter are:

- to introduce the idea of patterns;
- to explore how this idea may apply in the context of software maintenance;
- to suggest some practical ways of proceeding, along with some pitfalls to avoid; and
- to provide leads into the wider literature on the subject for the reader who wants to explore further.

The remainder of this chapter is structured as follows. In the next section we will briefly introduce the ideas of patterns, including the *process patterns* that will be particularly relevant to software maintenance. We will also discuss the closely related topic of *refactoring*. Then we discuss how these general techniques apply in the field of software maintenance, and in the following section we discuss how this application can be done in the context of an organization's processes. Finally, we discuss future trends and conclude.

BACKGROUND

Patterns

First, we give a little more detail concerning patterns, their origin in architecture, and their adoption in software design.

The term pattern in the sense meant here was coined by the architect Christopher Alexander. In books such as *A Pattern Language: Towns, Buildings, Construction* and *The Timeless Way of Building* (Alexander, 1979; Alexander et al., 1977) he collected solution parts that he had seen used successfully on several occasions, and wrote them up in a standard, easy to consult form. [A summary of the form used by Alexander can be seen in Figure 1, where all quotations are from Alexander et al. (1977, pp. x-xi and 833-837).] He explored the process of solving recurring problems by balancing different considerations (sometimes here, and often later, called *forces*) that anybody looking for a good solution in a constrained solution space faces. For example, in his pattern *Window Place,* he describes the tendency for people to want to be near the window in any room and points out that this may conflict with the desire to sit in a comfortable sitting area within the room. The solution suggested—which shares with many successful patterns an "obvious once you see it'' quality—is to make sure that there is a comfortable sitting area near the window. Individually, Alexander's patterns were not new insights; they were tried-and-tested solutions that had been used many times by different architects. Alexander's great contribution was to collect them together and to explain them and their connections in a standard way that made it easy for people to browse and consult his books. He intended that the patterns should form a *pattern language*; that they should be *generative* in the sense that once the designer has chosen a suitable subset of the available patterns, the design should in some sense emerge from them. Note that Alexander also insisted that one could use the same pattern many times without doing the same thing twice, so there is nothing mechanistic about the process of getting from a list of pattern names to a design.

Even at the time, Alexander was conscious of trying to introduce a significantly different approach to knowledge codification and transfer. However, in that field patterns never achieved the prominence that they now enjoy in software design. It would be interesting to study why that is; I suspect that it has to do, at least partly, with the different perceptions of individual creativity in the two fields.

Patterns in software design. It is in object-oriented software design that patterns have become a widely used and understood technique for recording and

Figure 1: Format of a pattern as used by Alexander.

1. A short, descriptive name for the pattern. May be adorned with one or two asterisks denoting increasing confidence in the pattern.
2. A picture showing "an archetypal example" of the pattern
3. A short paragraph on how the pattern "helps to complete certain large patterns"
4. A graphical divider (three diamonds)
5. A sentence (occasionally two) describing the problem, set in bold type. E.g., "Everybody loves window seats, bay windows, and big windows with low sills and comfortable chairs drawn up to them."
6. The main section, discussing what the pattern solution is and how, why, and when it works. May include photographs and/or diagrams.
7. Keyword "Therefore:"
8. A sentence (or two) summarizing the solution, again set in bold type. E.g., "In every room where you spend any length of time during the day, make at least one window into a 'window place.'"
9. A summary diagram.
10. A graphical divider (three diamonds).
11. A short paragraph on "smaller" patterns that may work with this pattern.

transferring problem-solving expertise. One pioneer was Erich Gamma, who later wrote with Richard Helm, Ralph Johnson and John Vlissides (1995) what is still the most widely used book of software design patterns; this is the book often known as the Gang of Four or GoF book. Another was Kent Beck, whose book (Beck, 1996) included patterns that recorded good habits of coding, as well as some concerning higher level design.

By comparison with Alexander's intentions for patterns, generativity is played down in the software design world. While the idea that software designs would be made entirely of patterns did appear—that is, some people believed (and may still believe) that there are object-oriented systems in which every class and every relationship between classes can be defined by the roles it plays in the patterns in which it takes part—this has never been a mainstream view. It is more common for a system to include the use of only a few patterns, deployed to solve particular problems that arise.

Is this simply because not enough patterns have been written down? Perhaps, but to answer this question we would have to understand what we mean by "enough." The object oriented design world was not short of methodologies before the arrival of patterns. There were plenty of books describing in general terms how to identify a good set of classes and define the interactions between objects of those classes. What was missing was more particular—descriptions of how to avoid the pitfalls that could occur even when following such a methodology, the pitfalls that experienced designers knew how to steer around. This was the gap that design patterns filled.

It is worth mentioning that we can already start to see the relevance of patterns to software maintenance. In many cases, the problems that design patterns solved were not so much problems of how to design a system in the first place, but of how to design a *maintainable* system. Patterns such as State, Strategy, Visitor (Gamma et al., 1995) and many others are concerned with localizing hotspots in the design, that is, ensuring that certain kinds of anticipated change would require minimal alterations to the system. Several formats have been used for recording design patterns. In general, the formats tend to be more structured than Alexander's, with a larger number of set parts; the pattern typically includes fragments of design diagrams, and often also fragments of code. Figure 2 describes the structure of the GoF patterns.

Whatever the format, a pattern is a structured, named description of a good solution to a common problem in context. Its name gives it a handle that can be used in conversation among people who are familiar with the pattern; people can say "Shall we use *Visitor*?" rather than having to explain a design at length. The use of an illustrative example helps the reader to understand; this is not to be confused with the examples of real use of a pattern that often occur later in the pattern description. (Ideally, a pattern should have been used at least three times by different organizations before being considered mature; this is the *Rule of Three*.) It includes a discussion of the pros and cons of the solution, helping the reader to consider

Figure 2: Format of a pattern as used by Gamma et al.

Name: Short; often descriptive, but at least memorable. Annotated with the name of the category into which this pattern falls, e.g., Object Behavioral.
Intent: A sentence or two summarising what the pattern achieves.
Also Known As: Aliases, since several design patterns are known under different names for historical reasons.
Motivation: The description of the overall problem; often includes a simple example.
Applicability: "Use the *Name* pattern when" and a list of pointers to its applicability, e.g., "many related classes differ only in their behavior."
Structure: One or more diagrams.
Participants: Brief description of the role played by each class in the pattern.
Collaborations: How the pattern works, dynamically.
Consequences: Description of the benefits of using the pattern where it is applicable, and of the drawbacks of doing so.
Implementation: Description of implementation issues, i.e., lower level considerations than were discussed under Consequences.
Sample code: With explanations; typically, the code corresponds to the simple example introduced under Motivation.
Known uses: Comments on real systems that used this pattern.
Related patterns: May include alternatives to this pattern, as well as "larger" and "smaller" patterns that may interact well with this one.

alternatives and choose the best way to proceed. In software design, a solution normally takes the form of a fragment of design; however, in software maintenance it is common for the most vital information to be about the *process* of arriving at a new situation, rather than about the new situation itself.

Process patterns. The term *process pattern* covers patterns in which the solution is not a fragment of an artifact, such as a design, but a way of carrying out a process. A variety of process pattern languages that, while not being specific to software maintenance, are relevant to this field have been considered.

For example, Alistair Cockburn has written a collection of patterns describing how to manage risks that arise in teamwork (Cockburn, 2001). His pattern *Sacrifice one person* applies in circumstances where members of the team are often interrupted by urgent tasks (for example, requests for progress reports, support on earlier products, etc.). Whereas it is natural—and in some circumstances optimal—to try to share out these interruptions so that no one member of the team gets unduly behind in their main task, Cockburn suggests that in certain circumstances it is better to designate one member of the team as the interface to the outside world and to have this person handle all interruptions so that the rest of the team may continue unhindered. Although the sacrificial victim's "real" work will have to carried by the rest of the team, this can be more efficient that allowing everyone to be interrupted. Some discussion by Cockburn and others follows, mentioning, for example, the need to be aware of the feelings of the person who is sacrificed. Other examples of pattern languages relevant but not exclusive to software maintenance include Brad Appleton's *Patterns for Conducting Process Improvement* (Appleton, 1997) and Scott Ambler's *Reuse Patterns and Antipatterns*. (An antipattern is a description of a bad way of doing something.)

A very useful web page collecting resources on process patterns is Ambler's *Process Patterns Resource Page* at http://www.ambysoft.com/processPatternsPage.html.[1]

Refactoring

Refactoring is a simple technique for making improvements in the design of an existing body of code, while minimizing the chances of introducing new bugs in the process. It has come to prominence recently in the context of *extreme programming* (XP); see for example Beck (2000).

As with patterns, refactoring makes explicit good practice that has been developed many times over by individual developers, rather than creating a radical new technique. It applies particularly when one attempts to add a feature to a system whose design is not well adapted to support that feature, whether that happens in initial development of the system or in perfective maintenance. The natural, naive approaches are either to ignore the deficiencies of the design and hack the feature in somehow, or, at the next level of professionalism, to improve the design at the same time as adding the feature. The idea of refactoring is to observe that the latter procedure is very error prone. One often—even normally—finds that the new feature does not work as intended, and then faces the task of determining whether it is the understanding of the redesign or the understanding of the new feature that is at fault. Worse is the case where the new feature works but regression testing shows that something old is now broken. Is the feature responsible in some interesting way for breaking something that previously worked, or is it just that the redesign has not been properly conceived or implemented? Complex debugging tasks like these can easily lead developers back to the original "hack it in somehow" solution; symptoms include people saying things like "I daren't touch this code, anything I do to it breaks it."

More experienced developers will recognize that the problem is not attempting to redesign parts of the underlying system, as such, but rather, trying to do too much at once, and especially trying to make changes to a system without a completely clear understanding of what the resulting behavior should be. Refactoring captures this idea. A refactoring is a change, usually a small change, made to a system *that is not intended to change its behavior;* that is, it is a change that contributes to improving the design of a system, for example, getting the system to a point where its design supports a new feature smoothly and easily.

This tends to be a counterintuitive idea for new developers, who, especially under time pressure, often consider that they haven't time to change the design of a system at all, let alone to do so in a succession of small steps, compiling and testing after each step. It is true that to work this way requires a supportive infrastructure of useful tests and usually the ability to compile and test a subsystem separately from the rest of the system; it will not work in a situation where the only way to test your code is to build an entire system that takes many hours to compile! Usually, however, the impression that this will be an inefficient way to work proves to be an illusion. It is human nature to discount the possibility that one will make mistakes, and so people tend to see the extra time that repeated compilations and test running will take and not see the extra time that the alternative, more error-prone procedure will take in debugging and bugfixing.

The general idea of refactoring, like the general idea of using design patterns, applies to any system in any paradigm. Again, however, it is in the object-oriented world that the idea has come to prominence. Martin Fowler has written a book (Fowler, 1999) that describes some of the commonest design improvements that can be made to object-oriented systems. He writes them up in a structured form reminiscent of that used for design patterns, describing explicitly what steps should be performed, in what order, and at what stages to run regression tests.

PATTERNS FOR SOFTWARE MAINTENANCE

In this section we consider the ways in which patterns may be relevant to the software maintenance issue, and what the major problems are, before going on in the next section to discuss practical ways of introducing their use into an organization.

Let us bring together the ways we have mentioned in which patterns and the related technique of refactoring may be relevant to software maintenance, in the sense that it would be useful for someone engaged in software maintenance to know something about them. They do overlap, but categorization seems useful at this stage.

1. A *design pattern* may have been used in the original development of a system for some reason, e.g., to facilitate changing something that the original designer expected to change.
2. It may be useful to introduce a *design pattern* by refactoring before changing the functionality of a system. Of course, this applies particularly where it later emerges that flexibility is needed beyond what was anticipated by the original designers.
3. *Process patterns* such as those in Cockburn (2001) and Ambler (1998a, 1998b) may be useful within a maintenance team, as elsewhere. Indeed, Ambler (1998a) includes a chapter on maintenance and support, but it is short and rather basic.
4. *Reengineering patterns*, a particular kind of process pattern targeted at legacy system problems, may be worth considering, particularly when a system needs strategic redesign that is not easy to tackle as a succession of refactoring steps.
5. *Pattern-writing* may be a useful way to focus on lessons learned within an organization at a project review meeting.

Process patterns, as discussed above, are also covered in the "Background" section, and for further information the reader is referred to the references cited. We shall discuss design patterns, as most of the available literature focuses on design patterns as used by an original development team, rather than by a maintenance team. We will briefly discuss reengineering patterns, referring out to the literature, and will spend most time on pattern-writing which has not received so much attention in the patterns literature.

Issues, Controversies, Problems

Before going on, however, it seems wise to touch briefly on some of the reasons why patterns are not a "silver bullet," and on some of the controversies that arise inside and outside the patterns community.

Do patterns work at all for knowledge reuse? On the face of it the answer to this question has to be a resounding Yes, given the enormous popularity of patterns in object-oriented software design as evidenced by the market for books and training courses in particular. Yet there is a school of thought that patterns are no more than the latest buzzword, which will fade as quickly as it came, and this view can plausibly claim support from the field of organization theory, in which the severe difficulties in knowledge articulation and transfer are well understood. A full discussion of the issues is beyond the scope of this chapter (see d'Adderio, Dewar, Lloyd, & Stevens, 2002). Here it suffices to say that while we would not wish to argue against the evidence of our eyes that many people and organizations do find patterns immensely useful, it is important to remember that reading a pattern will not—on its own—enable a novice to solve a problem as an expert would, even if an appropriate pattern is available. A pattern provides a bare skeleton of what is needed, which should be helpful, but expertise in the field is still needed to put the pattern into practice. Experienced people are in no danger of being made redundant by pattern catalogues! In fact, direct knowledge transfer may not even be the most important function of patterns; it may be that the thought-provoking nature of patterns is even more important.

Is it realistic to have cross-organizational patterns for software maintenance? It is arguable that identifying and communicating expertise in software maintenance is a more important and serious problem than it is in software design. One reason for this is that the existing system and business environment has a huge effect on the available solutions and their relative merits. Therefore, whereas in object-oriented software design the usual way to use patterns is to consult one of several large catalogues of "general purpose" design patterns, in software maintenance it may be more effective for an organization to build up its own catalogue of patterns, adapted to its own situation; that is, an internal pattern catalogue may be used as a repository of organizational knowledge. In the next section we will suggest how the development, maintenance, and use of such a catalogue may be built into the organization's software maintenance cycle in a cost effective way.

Use and Abuse of Design Patterns

In this section we consider how design patterns, that is, patterns that describe design structures that are useful in particular circumstances, may influence software maintenance. Because today most of the existing software design patterns are object-oriented, this section is more relevant to maintainers working with object oriented software than it is to others. However, there is nothing inherently object oriented about design patterns: successful designs in any paradigm could be recorded in pattern form.

The contribution of the original developers. Software maintainers' jobs may be made significantly easier or immeasurably harder by the way in which the original developers used design patterns. Used properly, design patterns can contribute to a flexible, clean design in which code duplication is minimized, structure is clear, and anticipated modifications are straightforward. Used badly, they can result in a morass of incomprehensible code with over-complex dependency structures.

The essential thing to understand is that most design patterns are inherently complex, non-obvious solutions. That is why they are worth writing down in the first place; they are not what a designer will arrive at by following his or her nose. Indeed, when I teach design patterns I generally begin by presenting a problem and asking students to come up with an initial, straightforward design, before we study an applicable pattern. Whether the students' initial design or the pattern is the better solution depends on the real circumstances of the design problem.

To illustrate, let us consider one of the simpler Gang of Four patterns, Strategy. The intention of this pattern is to make it straightforward to apply different algorithms to the same problem, depending on the circumstances that prevail at run time.

A class (which, following Gamma et al. [1995], we will call Context for purposes of exposition) has been allocated various responsibilities, one or more of which involve applying the algorithm in question. It is required to separate the choice of algorithm from the invocation of the algorithm, for example, so that one part of the code can deal with algorithm selection and the rest can interact with the Context without needing to know which algorithm is in use. How the algorithms are swapped, however, is the concern of Context only; the outside world will not know.

A naive solution will involve Context storing some flag or codeword representing the algorithm currently in use, and appropriate methods of Context will involve if/else or case statements that branch on that codeword and execute the relevant algorithm. If the number of algorithms is large, or if the structure of shared and non-shared code is complex, the code of Context will become hard to read and maintain.

The Strategy pattern shows how to split the code into one coordinating class, one abstract class, and one class for each concrete algorithm, or *strategy*, in use.

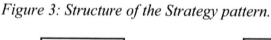

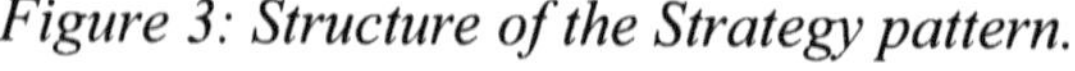
Figure 3: Structure of the Strategy pattern.

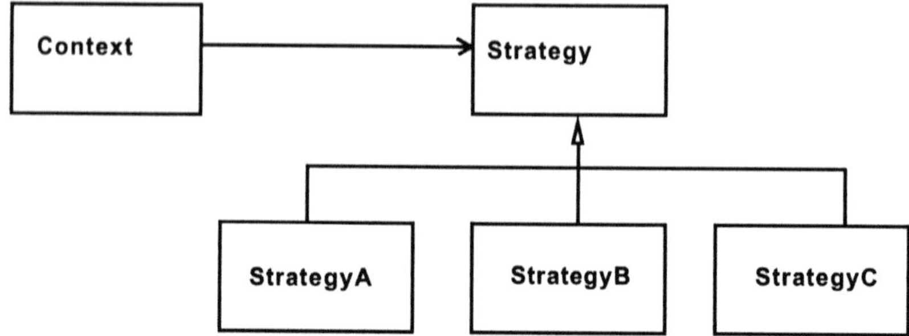

The new Context class contains an attribute that is a reference to an object of some subclass of Strategy, and provides code for switching algorithms. The new abstract Strategy class defines the interface that all algorithms are required to satisfy, and may also define common supporting methods or data. The specific subclasses of Strategy implement the unique parts of each algorithm. Figure 3 illustrates.

The result is that, although the class structure of the resulting system is more complex, it is easier to find the relevant pieces of code for a given purpose. More important for our purpose here, it is straightforward to add, remove, or modify an algorithm during maintenance; for example, adding a new algorithm requires creating a new subclass of Strategy and modifying the Context code that selects an object of that class as the current strategy.

Whether the additional structural complexity is worth accepting depends on the situation. It is more likely to be acceptable if there are expected to be many algorithms or if they have a complex structure of sharing pieces of code. Even in such an apparently technical problem, non-technical factors may be important. For example, if several people are to be involved in implementing the functionality of Context, it is far more likely to be worth using the Strategy pattern than if the whole is to be implemented by one developer, because splitting up the functionality in a clear, easy to understand way will then be important.

Please note that this description has been extremely abbreviated; the full discussion in Gamma et al. (1995) occupies nine pages.

An anecdote told by Michael Karasick when he presented his very interesting paper (Karasick, 1998) at a recent Foundations of Software Engineering conference is also pertinent. The group was working on a development environment that was intended to be long lived and very flexible; it was intended that a variety of tools to perform different kinds of analysis and manipulation of the structure of the user's program would be added in future. This is, in classical design patterns terms, an obvious situation in which to be using the Visitor pattern. However, the elegant solution proposed by this pattern involves very complex interactions and is notoriously difficult for people to "get their heads around." Moreover, the development team had a high rate of turnover; new recruits had to be constantly educated in the architecture of the system. In the end, this was the major factor that convinced the team that it was not sensible to use Visitor. They used instead a simpler approach that involved the use of run-time type information. For them, the drawbacks of this solution—for example, the possibility of type errors in the tool's code that the compiler would not be able to detect—were acceptable, given the difficulty of making sure that everybody who needed to be was happy with the Visitor solution. This is not, of course, to say that another group or the same group at a different time would have come to the same conclusion. For example, design patterns are much more widely used and taught now than they were then. If the probability that new recruits would already be familiar with Visitor had been a little higher, the IBM team's decision might have been different.

Whether and how to introduce a design pattern. So, if a maintenance team inherits a system that does not make use of a design pattern in what seems like an obvious place for one, should it introduce the pattern? And are the questions to be answered the same as those for the original development team, or are they different in character?

The maintenance team may be better placed than the original development team was to see what flexibility the design needs. The real requirements on a system are generally much clearer once the system is in use; whereas the original development team may have guessed that a certain kind of flexibility would be required in future, the maintenance team may be in a position to know that it will be. If there are several change requests in the pipeline that would be easier to fulfill if a particular design pattern were in place, then it may be sensible to adopt the design pattern when the first such change request is actioned.

Let us return to our Strategy example. Suppose that the system currently incorporates two different algorithms, and that a boolean attribute of a monolithic context class ParagraphSplitter controls which one is in use. Several methods of this class have the form

```
if(usingAlgorithmB) {
// do algorithm B stuff
...
} else {
// do algorithm A stuff
}
```

The change request being considered is to add a third algorithm, C.

When only two algorithms were available, Strategy, if it was considered at all, probably seemed like overkill. This judgment may or may not have been "correct"; in any case, it was an educated guess based on judgments about exactly such unknowns as whether it would be necessary to add new algorithms in future.

The maintainer has several options:

1. Keep usingAlgorithmB as it is, add another boolean usingAlgorithmC, and insert a new if/else statement inside the existing else case.
2. Replace the boolean usingAlgorithmB by a variable, say algorithm, of an enumerated type A,B,C, and replace the if/else by a three-way case statement.
3. Use Strategy; that is, split off most of ParagraphSplitter's functionality into new classes for each of the three algorithms, together with an abstract algorithm class.

The first is likely to be a mistake; it fails to make explicit the invariant that at most one of the booleans usingAlgorithmB and usingAlgorithmC should be true, and the code structure will be confusing. (The exceptional case where this might be acceptable would be when algorithm C was in fact a variant of algorithm B, so that the branch "Are we using a B-type algorithm? If yes, then is it, more specifically, algorithm C?'' would reflect the domain.) To decide between the last two options, we would need to know more about the problem, for example, how complicated the algorithms are. A rule of thumb is that case statements are unlikely to be confusing if the whole of the case statement can be seen on one screen, so that in such a case, the second option might be reasonable.

If the third option is chosen, the best way to introduce the Strategy pattern is by a sequence of refactorings. First, the system should be changed so that it uses the Strategy pattern with only the two existing algorithms. That can itself be done in several stages; most obviously, one could first separate out a new Strategy class from ParagaphSplitter, moving the flag usingAlgorithmB and the code for the two algorithms into that class, and creating an appropriate interface for ParagraphSplitter to use to communicate with its Strategy. Then the Strategy class itself can be split into an abstract base class with common utility methods, and two concrete subclasses AlgorithmA and AlgorithmB, each implementing the algorithm interface agreed upon at the previous stage. Most care is needed in deciding how the creation and swapping of the Strategy objects is achieved, since this will introduce some dependencies on the concrete subclasses of Strategy and these should be localized and limited. Whichever resolution of that problem is used, the system should be back in working order, passing its regression tests, before the new algorithm is added; this separates concerns about the new algorithm from more general concerns about the architecture of the system.

Software Reengineering Patterns

Software reengineering patterns have been investigated by several research groups, at least two of which (the groups at Edinburgh and at Bern) coined the term independently before learning of one another's work. The Edinburgh group, with which I work, (funded by the grant *Software Reengineering Patterns* from the UK Engineering and Physical Sciences Research Council) includes not only people whose backgrounds are in computer science and software engineering, but also a management scientist, a mechanical engineer, and an organizational theorist. Naturally, we take an interdisciplinary approach to the subject. The work of the Software Composition Group at the University of Bern is more technical in nature, having focused primarily on patterns for restructuring parts of object-oriented designs. Our patterns so far are preliminary and few in number, but may provide at least a starting point and some initial record of best practice. For example, the pattern *Deprecation* discusses the important, common, but often overlooked problem of how to manage the evolution of a library (of functions, components, or whatever) that is being widely used. The basic technique is that when elements must be dropped from the library, they are as a first step labelled "deprecated" before being dropped in a later release of the library. There are, however, several considerations—such as the nature of the applications that use the library, the willingness of the library's users to upgrade, etc.—that affect the success of such a strategy and which are easily overlooked.

For more information, see the websites of the groups at http://www.reengineering.ed.ac.uk and http://www.iam.unibe.ch/~famoos/patterns/ respectively.

Anthony Lauder and Stuart Kent at the University of Kent have pursued a program of work grounded in Lauder's work over several years with Electronic Data Processing PLC, a software company specializing in software for sales and distribution. Lauder worked with EDP to develop parallel catalogues of *petrifying patterns* and *productive patterns*. Essentially, this approach separates consideration of commonly occurring problems from discussion of tried-and-tested solutions. "Petrifying pattern" is Lauder's coinage. The term captures both the tendency of unsuccessful practices to reduce flexibility, turning parts of a system to stone, and the effect on the unprepared developer of having to deal with the results of such practices! Petrifying patterns are not patterns in the classical sense, but, they can be illuminating especially when paired with productive patterns. Similar descriptions of problems have elsewhere been called anti-patterns.

As an example, we will consider the petrifying pattern *Tower of Babel* and its antidote productive pattern *Babel Fish*, both from Lauder (2001). As before, the reader should be aware that what follows are necessarily summaries of much longer descriptions.

Tower of Babel. The problem described in this pattern is the difficulty of making applications written in different languages interoperate. This can, for example, lock developers into a legacy system. If new functionality needs to be added to the legacy system and the developers do not perceive that there is any easy way of making the legacy system interact with a new, separate component that provides the new functionality, they are driven to implement the functionality inside the legacy system, perhaps in a legacy language. Conversely, migration of functionality out of a legacy system is also hampered. One of the concrete examples given by Lauder of how this problem arose in EDP concerns its sales order processing system that interacted with its legacy financials package. EDP wanted to replace the legacy financials package with a better third-party financials package, but found this impractical given that the systems were in different languages (Commercial Basic,

Visual C++), incorporating different assumptions about implementation technology.

Babel Fish. The solution described by Lauder is to localize the inter-application communication into two-way gateways external to each application, called Babel Fish. A Babel Fish allows its application to send messages to other applications and receive messages from them, without concern for the language translation necessary for communication; translation to and from a language-independent form is performed by the Babel Fish. A single Babel Listener routes these language independent messages, using a configuration file to determine a translation from symbolic names for applications to their precise addresses on a network. EDP has developed a Babel Fish Framework to ease the development of Babel Fish for new applications.

The pattern discusses the impact of this strategy on systems maintenance within EDP. Principally, the technique has been used to ease the incremental migration of legacy systems away from legacy languages such as COBOL and Basic. The existence of a well-understood way to implement interaction between a modern replacement and a legacy application has apparently enabled EDP to migrate functionality gradually, because applications are effectively shielded from the progressive migration of applications to which they talk; the translation from symbolic names to actual addresses, and even the language of the migrating application, can change without affecting its clients. The pattern also discusses alternatives such as ad hoc connections between specific applications, and the use of CORBA.

It is interesting to consider to what extent patterns such as these are useful outside their originating organization. Lauder stresses the importance of a maintenance team building up its own catalogues of patterns, and his patterns were built up within EDP for EDP's own use, although their presentation suggests also an intention that they may be used elsewhere. It is arguable, for example, that the details of Babel Fish are unlikely to be of interest to another organization, at least in the absence of a freely available Babel Fish Framework, since to develop and use such a framework from scratch would probably not be significantly easier than adopting, for example, CORBA. Indeed, the pattern does not go into sufficient detail for the external reader to get more than a broad impression of what is intended, and it is probable that the inclusion of such details would make the pattern either much less widely applicable (as it codified assumptions that apply within EDP but not elsewhere) or much more unwieldy (as it proposed mechanisms for avoiding such assumptions, approaching the generality of something like CORBA). Within EDP, on the other hand, the pattern will have been useful as a codification of solutions already in use, capitalizing on in-house software and expertise already available to provide support and examples.

Currently the best source of information on Lauder's work is his recent PhD thesis (Lauder, 2001); earlier work in the program is reported by Lauder and Kent (2000, 2001).

Using Patterns to Capture and Enhance an Organization's Expertise

Thus far we have focused on using patterns that have already been written by others. However, earlier we mentioned as a possible impediment to this approach, the possibility that solutions to some important software maintenance problems may be necessarily specific to a given organization. The following fictionalized example illustrates.

Organization A has determined the strategy of reducing and eventually eliminating its dependence on a large transaction-based legacy system. Team B is the maintenance team charged with carrying out this strategy; whenever changes are required to a transaction within the legacy system, the opportunity is taken to migrate the functionality provided by that transaction and any related ones to an appropriate growing system. Over some years, Team B accumulates considerable understanding of how to do this. Its knowledge includes understanding which transactions should be moved as a whole, and to which system; the team knows how to proceed to make the necessary code and database changes; how to update the user interfaces of both source and target system. It has also accumulated expertise in how to roll the change out to the organization, how new releases are made and coordinated, who needs to be trained and how, and what support is needed for the modified system.

Little of this expertise is likely to be recorded in any generic way. There may possibly be documents recording the plan of action for carrying out *particular instances* of this migration, but a new member of Team B will still need to have the relevance of such documents explained by an "old hand;" even then, there will be considerable gaps. The team will collectively understand under what circumstances particular parts of a solution do or do not apply, and what variants are permissible. Provided that the team is sufficiently stable, it may be that allowing new team members to learn from old team members is a sensible and efficient way to proceed. It has its dangers, however; in times of high turnover it may begin to break down, and if more drastic action such as completely outsourcing the maintenance of the system taken, this understanding is likely to be lost.

Under such circumstances Team B might consider it useful to write a pattern to describe its collective knowledge. It would need to discuss the common features of the cases of the problem it has experienced, and choose an appropriate form, either inventing its own or adopting one of the many formats that have already been used. Probably it would be most effective to develop the pattern iteratively, as discussed in the next section, but eventually, it could build up a pattern or, if the problem is better split into parts, a pattern language that might be useful to new members of the team. What might not be so obvious at the outset is that there seem to be benefits even to experienced team members in attempting to codify their knowledge in this way; for example, interesting discussions are likely to arise out of considering why a particular instance of the problem was solved in a different way from most.

Most software development organizations routinely make use of the Web, and it is natural to use this medium for dissemination of the codified expertise, whether in pattern form or some other. This has many advantages, such as its being easy to produce patterns using standard HTML authoring tools, ease of version control (compared with paper-based solutions), and familiarity of the interface. There are a number of disadvantages as well: while it is easy for an individual to search purposefully for a particular pattern in a web-based tool, it is less easy for a team meeting to consult its pattern collection during its discussions, and people will be unable to use patterns as their bedtime reading. Experimenting with providing a "printer-friendly" version (with version information) might be worthwhile.

PATTERNS INTO PRACTICE

Considering how and why patterns can in principle be written and used is only a small part of what has to be done, however. At least as important is to consider whether and how pattern use can be turned into an integral part of the software evolution process within an organization.

Throughout this section, the reader is invited to consider how the discussions fit with his or her own approach to process quality and maturity. See also the chapter on *Software Maintenance Maturity* in this volume.

Recording the use of a design pattern. Whether a design pattern was used in the initial design of a system or whether it has been added during the maintenance process, it is important that future maintainers of the system understand the reasons why the design is as it is; otherwise, not only will they have a hard time comprehending the system, but also, they may unwittingly break the pattern use. If this happens, the system is left with the disadvantages both of using and of not using the pattern: the design retains the comparative complexity that it was given by the pattern use, but it does not retain the benefits of the now-broken pattern!

There are certain obvious measures that can be taken, such as giving standard names to the participants in the pattern, including appropriate comments in the code, and so on. If design diagrams for the system are available, then there is the further possibility of using these to record pattern use. The Unified Modeling Language (UML)(OMG, 2001) even provides a standard notation to record the use of patterns, in which the pattern is shown as an oval and visually related to the classes that play the standard roles in the pattern. Figures 4 and 5 illustrate two variants of the notation in the case of the Strategy pattern example; for more details, see OMG (2001) or a good UML book.

Learning a repertoire of patterns. To a large extent, becoming familiar with collections of patterns is an essentially individual activity; little seems to be gained by attempting to mandate the use of patterns. Managers can facilitate the process by purchasing books of patterns and perhaps by putting items such as "any relevant patterns" onto meeting agendas and document templates. For example, a team could routinely ask whether any pattern is applicable when it decides how to tackle a difficult modification request. It is to be expected that the answer will frequently be that the team does *not* know any relevant pattern; once the problem is solved, if the solution seems generally applicable, the team might consider summarizing the solution it found in pattern form.

Many pattern-reading groups exist; typically, the group meets for lunch or over a drink to discuss one pattern or a small number of patterns. Teams that are

Figure 4: UML's pattern notation, variant 1.

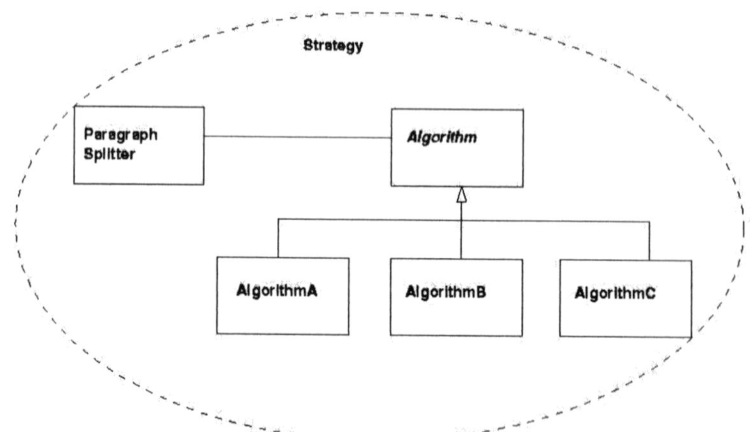

Figure 5: UML's pattern notation, variant 2.

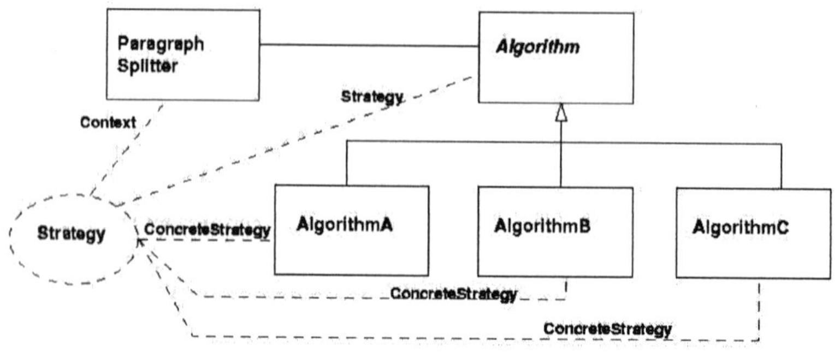

interested in getting into patterns might consider starting their own small group. There are definite advantages in a team of people being familiar with the *same* collection of patterns, because then the pattern names can be used in conversation as a kind of high-level (system or process) design vocabulary.

Writing patterns. It may well be impossible to identify dedicated effort for writing patterns, and integrating pattern development and use into the process organically may be the only way to proceed. Even if it is possible to allocate effort specifically to pattern writing, an organization's pattern catalogue needs to be a living artifact if it is to succeed; a one-off effort may help to start the process, but it will not suffice on its own.

Therefore, an organization that wishes to write patterns needs to identify the existing opportunities for doing so. For example, if there are regular project meetings to assess progress and discuss problems, then perhaps pattern writing should be a standing agenda item. Writing good patterns is not easy and is an art rather than a science, so it should not be expected that a pattern will be produced in its final form at such a meeting. There are several "pattern languages for pattern writing" available on the Web that may help a team new to pattern writing to get started; in book form, John Vlissides' short book (Vlissides, 1998) is well worth reading for this purpose, even though it concentrates on design patterns rather than process patterns. It is important to remember that, provided that there is some way of identifying how much or little reliance can safely be put on a pattern (Alexander's asterisks, for example) it does not matter if a particular pattern is just a rough draft; if it proves useful but incomplete, it can be improved at a later date. In such cases it is especially important that readers of the draft pattern know whom to contact for more information and discussion.

Notice that patterns written in this way need not be organization specific. Solutions that are gleaned from other sources such as books, papers, or conversations might be recorded in this way for the benefit of team members who are not familiar with the original source. In such a context it may be sensible to write very abbreviated patterns, that might not describe the solution in detail, but that would, rather, provide a pointer to the full source, such as "see pages 73-78 of Bloggs' book *Software Maintenance* in the team room."

FUTURE TRENDS

What are the Next Steps for Pattern Use in Software Maintenance?

Design patterns and process patterns that are not specific to software maintenance will continue to be written; it is to be hoped that their visibility to all those who may find them useful will continue to increase. In my view, one of the signs of the increasing maturity of the patterns world is the gradual diminution of the "hype" surrounding patterns. At this stage, it is important that we investigate the limits of what patterns can do and understand where they are and are not appropriate. Part of this investigation involves an understanding of what it is that people get out of patterns. I have hinted at a belief that patterns are not, in fact, mainly used as cookbook solutions to current problems. I suspect that patterns have several more important roles, for example, as learning material and as thought experiments.

Patterns specific to software maintenance or, yet more specifically, to software reengineering have a more uncertain future. My own experience leads me to suspect that the use of such patterns by a maintenance organization to document the solutions that work in its own context will ultimately prove to be more fruitful than the documentation of solutions that apply to a wide class of organizations. It seems to me that maintenance is so organization-dependent that it is difficult to codify, in pattern form or in any other way, much that can genuinely be of use in many organizations.

On a cynical note, one might also argue that the use of a fashionable technique such as patterns might help to attract more attention to the Cinderella discipline of software maintenance.

CONCLUSIONS

In this chapter we have introduced the idea of using patterns in software maintenance. We have covered three main types of pattern use: the use of design patterns, the use of existing process patterns, and the writing and use of organization-specific patterns. We have briefly discussed some practical ways in which patterns of these different kinds may be relevant, and how they can be introduced into a maintenance organization's practice.

A chapter such as this can only be an introduction, but the references that follow and the URLs in the text should provide leads into the wider literature for any reader who would like to explore further.

ENDNOTES

[1] This, like all URLs in this chapter, was last checked on 1 December 2001.

REFERENCES

Alexander, C. (1979). *The Timeless Way of Building*. Oxford, UK: Oxford University Press.

Alexander, C., Ishikawa, S., Silverstein, M., Jacobson, M., Fiksdahl-King, I., & Angel,

S. (1977). *A Pattern Language: Towns, Buildings, Construction.* Oxford, UK: Oxford University Press.

Ambler, S. (1998a). *More Process Patterns: Delivering Large-Scale Systems using Object Technology*. Cambridge, UK: SIGS Books/Cambridge University Press.

Ambler, S. (1998b). *Process Patterns: Building Large-Scale Systems using Object Technology*. SIGS Books/Cambridge University Press.

Appleton, B. (1997). Patterns for conducting process improvement. http://www.enteract.com/~bradapp/docs/i-spi/plop97.html.

Beck, K. (1996). *Smalltalk Best Practice Patterns*. New York: Prentice Hall.

Beck, K. (2000). *Extreme Programming Explained: Embrace Change.* Englewood Cliffs, NJ: Addison-Wesley.

Cockburn, A. (2001). The risk management catalog. http://members.aol.com/acockburn/riskcata/risktoc.htm.

d'Adderio, L., Dewar, R., Lloyd, A., & Stevens, P. (2002). Has the pattern emperor any clothes? A controversy in three acts. *Software Engineering Notes, 27*(1), 31–35.

Fowler, M. (1999). *Refactoring : Improving the Design of Existing Code*. Reading, MA: Addison-Wesley.

Gamma, E., Helm, R., Johnson, R., & Vlissides, J. (1995). *Design Patterns*. Reading, MA: Addison Wesley.

Karasick, M. (1998). The architecture of Montana: An open and extensible programming environment with an incremental C++ compiler. In *Proceedings of the ACM Sigsoft Sixth International Symposium on the Foundations of Software Engineering* (pp. 131–142).

Lauder, A. (2001). A productive response to legacy system petrification. PhD thesis, University of Kent at Canterbury.

Lauder, A. & Kent, S. (2000). Legacy system anti-patterns and a pattern-oriented migration response. In P. Henderson (Ed.), *Systems Engineering for Business Process Change* New York: Springer-Verlag.

Lauder, A. & Kent, S. (2001). More legacy system patterns. In P. Henderson (Ed.), *Systems Engineering for Business Process Change*, (2). Berlin: Springer-Verlag.

OMG (2001). Unified Modeling Language Specification Version 1.4. OMG formal/2001 2001-03-67 document. Available at www.omg.org.

Vlissides, J. (1998). Pattern Hatching: Design Patterns Applied. Reading, MA: Addison-Wesley.

Chapter V

Enhancing Software Maintainability by Unifying and Integrating Standards

William C. Chu
Tunghai University, Taiwan

Chih-Hung Chang, Chih-Wei Lu, and Yeh-Ching Chung
Feng Chia University, Taiwan

Hongji Yang and Bing Qiao
De Montfort University, England

Hewijin Christine Jiau
National Cheng Kung University, Taiwan

Software standards are highly recommended because they promise faster and more efficient ways for software development with proven techniques and standard notations. Designers who adopt standards like UML and design patterns to construct models and designs in the processes of development suffer from a lack of communication and integration of various models and designs. Also, the problem of implicit inconsistency caused by making changes to components of the models and designs will significantly increase the cost and error for the process of maintenance. In this chapter, an XML-based unified model is proposed to help to solve the problems and to improve both software development and maintenance through unification and integration.

INTRODUCTION

Software systems need to be fast time-to-market, evolutional, interoperable, reusable, cross-platform, and much more. Maintaining software systems is now facing more challenges than ever before, due to 1) the rapid changes of hardware platforms, such as PDA, WAP phone, Information Appliance, etc., 2) emerging software technologies, such as Object Oriented, Java, middleware, groupware, etc., and 3) new services, such as E-commerce, mobile commerce, services for Application Service Provider (ASP), services for Internet Content Provider (ICP), etc.

Due to the high complexity of software systems, development and maintenance usually involve teamwork and high cost. However, most systems are developed in an ad hoc manner with very limited standard enforcement, which makes the software maintenance very difficult. De facto standards, such as Unified Modeling Language (UML) (OMG, 2001), or XML (Connolly, 2001; Lear, 1999), are used to reduce communication expenses during the software life cycle and to increase maintainability and reusability. Design Patterns (Gamma, Helm, Johnson, & Vlissides, 1995) are reusable solutions to recurring problems that occur during software development (Booch, 1991; Chu, Lu, Yang, & He, 2000; Holland, 1993; Johnson & Foote, 1988; Lano & Malik, 1997; Meyer, 1990; Ossher, Kaplan, Harrison, Katz, & Kruskal, 1995; Paul & Prakash, 1994; Xiao, 1994). However, these standards usually only cover partial phases of the software process. For example, UML provides standard notation for modeling software analysis and design. But lacking of support in the implementation and maintenance phases, design patterns offer help to the design phase, and component-based technologies focus on the implementation phase.

In other words, these standards are not talking to each other currently, and therefore designers need to spend a lot of manual effort mapping and integrating these standards while crossing each phase of the software life cycle. The activities of software maintenance involve the whole software life cycle, including requirement, design, implementation, testing, and maintenance phases. Not only the model used in each phase, but also the mapping and integration of models between phases will affect the efficiency of software maintenance.

Without unifying and integrating these standards, the consistency of the models cannot be maintained, and the extent of automation is very narrow. This chapter proposes an XML-based meta-model to unify and integrate these well-accepted standards in order to improve maintainability of the software systems.

This chapter will discuss the adopted standards, including UML, design patterns (Gamma et al., 1995), component-based framework, and XML. A comparison and mapping of these standards will be presented. An XML-based unified model is used to unify and integrate these various models.

The remaining sections of the chapter are organized as follows. Background information is introduced below and the following section proposes a new approach as XML-based Unified Meta-Model (XUMM) to unifying and integrating existing software development standards. Then, examples and experiments are given and finally future trends and conclusions will be discussed.

BACKGROUND

Developing and maintaining a software system are very difficult tasks due to their high complexity. Various models and standards have been proposed to reduce the complexity and the cost of software development. However, no standard or model can cover all dimensions of the software process.

In order to understand the activities of software development and maintenance, we should first understand the process of the software life cycle. Figure 1 shows a typical process of the software life cycle, which includes analysis, design, implementation, and maintenance phases (Sommerville, 1996).

During the process of system development, various kinds of domain knowledge, methodologies, and modeling standards will be adopted according to the specific requirements and individual expertise. As a result, the difference between methodologies and standards causes the development/maintenance to be difficult, and information can be lost while transforming from one phase to another. In the following, related methodologies, software standards, and studies are surveyed to disclose the problem itself, as well as some noteworthy efforts responding to that demand.

Object-Oriented Technology and UML

Object-oriented (OO) technology is a remarkable software methodology that organizes data in ways that "echo" how things appear, behave, and interact with each other in the real world. OO technologies greatly influence software development and maintenance through faster development, cost savings and quality improvement (Rine, 1997; Wirfs-Brock & Johnson, 1990). Object-Oriented Analysis and Design (OOA/D) (Booch, 1994) follows the concept of OO technology and thus has become a major trend for modern software development and system modeling.

Figure 1. A typical process of the software life cycle.

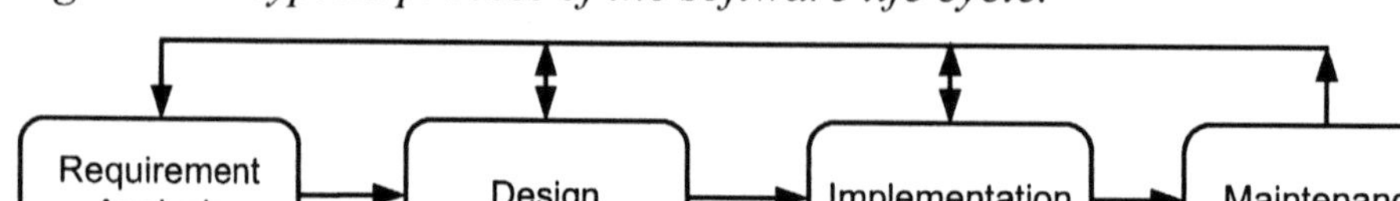

A sign of the maturity is the convergence of object-oriented modeling notations in the form of the Unified Modeling Language (UML) (OMG, 2001). UML is used to specify, visualize, construct, and document the artifacts of software systems. UML defines the following graphic and notations to express important domain-related concepts: *use case diagram, class diagram, behavior diagrams* (such as collaboration diagram) and *implementation diagrams* (such as component diagram). Each diagram of UML contributes to the understanding of system analysis and design. UML is rapidly becoming the first choice for object-oriented modeling in general. However, the lack of formality in UML prevents the evaluation of completeness, consistency, and content in requirements and design specifications (Bourdeau & Cheng, 1995).

Formalization of OOA/D

A formal method is a systematical and/or mathematical approach to software development; it begins with the construction of a formal specification describing the system under development (Bourdeau & Cheng, 1995). Formal specifications document software requirements or software designs using a formal language. A formal specification can be rigorously manipulated to allow the designer to assess the consistency, completeness, and robustness of a design before it is implemented. The advantages of using formal methods are obvious in that the uses of notations are both precise and verifiable, and the facilitation of automated processing is feasible. Much progress was made in formal methods in both theory and practice in the early of 1990s, most notably, the development of the Z notation (Spivey, 1992). Z notation models the data of a system using mathematical data types and describes the effects of system operations using predicate logic and the availability of supporting tools such as ZTC (Jia, 1998b) and ZANS (Jia, 1998a). To combine the strengths of object-orientation and formal methods seems. However, many aspects of object-oriented analysis and design models still remain informal or semi-formal, such as the data and operation specifications (Jia, 1997). The informal methods and approaches are seductive, inviting their use to enable the rapid construction of a system model using intuitive graphics and user-friendly languages; yet they are often ambiguous, resulting in diagrams that are easily misinterpreted (Cheng, Campbell, & Wang, 2000).

Some researchers tried to formalize and enable developers to construct object-oriented models of requirements and designs and then automatically generate formal specifications for diagrams (Cheng et al., 2000). Other projects have explored the addition of formal syntax and/or semantics to informal modeling techniques (Hartrum & Bailor, 1994; Hayes & Coleman, 1991; Moreira & Clark, 1996; Shroff & France, 1997). In Jia and Skevoulis (1998), a prototype tool—Venus, is proposed to integrate the popular object-oriented modeling notation

UML and a popular formal notation Z (Jia, 1997). Wang and Cheng (1998) give a brief overview of the process that has been developed to support a systematic development of the object-oriented models from requirements to design. But such studies only give partial consideration to all the phases during development or only address issues in a specific domain or application.

Other Modeling/Developing Techniques and Standards

Design Patterns. These are recognizable traces or blocks found in a series of observations or in a system. We can observe familiar usages and constructions during a system design that describe a problem and its solution; these are called *Design Patterns* (Gamma et al., 1995). The objective of design patterns is to enable designers to reuse well-known and successful designs and architectures from experts' experience more easily. Design patterns help programmers choose design alternatives that make a system reusable and avoid alternatives that compromise reusability. Design patterns can even improve the documentation and maintenance of existing systems by furnishing an explicit specification of class and object interactions and their underlying intent. In short, design patterns help a designer get a design "right" faster (Gamma et al., 1995). Design patterns are one of the hottest topics related to software development by providing recommended concrete solutions (Yau & Dong, 2000); nevertheless, it takes into account much about the analysis phase and leaves few cues about the consistency and integration of software components.

In any domain, there are certain characteristic problems inside the design activity that occur repeatedly. It is obviously important if we can make such knowledge explicit and publicly available to other practitioners. The most important advantage by doing so it that the inexperienced designers can gain access to a library of techniques that are immediately applicable, thus making a good design a lasting one. This is especially critical for the process of developing enterprise in the relevant design technique. By the inheritance of object-oriented design technology, design patterns have been grabbed as the major way of codifying the ways in which expert designers tackle and solve particular commonly occurring problems in design.

Framework. Framework is a software technology that contains software component reuse and design reuse. High reusability is the result of using framework to develop the same domain software system on existing architecture (Moser & Nierstrasz, 1996). Framework is a set of abstract classes and concrete classes that collaborate with each other, and contain their own interface definition of instances (Chen & Chen, 1994). These classes are constructed with a reusable application architecture for some domain. A developer can inherit these instances of classes to construct a new system. Basically, framework provides an environment to support all the activities for software development; however, for most of the cases only specific domain applications with limited scale benefits.

Design patterns and framework are two approaches to solving the problem of enabling certain elements of designs to be recorded and reused, but they differ significantly. We will compare them in terms of what they provide and how they intend to be used. First, design patterns and framework differ in terms of size. Inside of a framework, more than one design pattern may be applicable. Second, design patterns and framework have different contents. A framework defines a complete architecture that is a mixture of design and code. But a design pattern is a pure design idea that can be adapted and implemented in various ways according to the choice of the developers. From these comparisons, we know design patterns are more portable than framework.

Component Based Software Engineering (CBSE). As stated in Szyperski and Pfister (1997), a *software component* is a unit of composition with contractually specified interfaces and explicit context dependencies only. A software component can be deployed independently and is subject to composition by third parties. By using the well-proven software parts, reusable component software engineering has the advantages of fast development, easy configuration, stable operation, and convenient evolution. Component-based software reuse is considered one of the most effective approaches to improve software productivity. The research of software reuse covers the scope of software components identification, extraction, representation, retrieval, adaptation, and integration. Only a few researchers (Chu et al., 2000) have addressed the whole process. In practice, component-based software engineering still suffers from some problems. Smith, Szyperski, Meyer, and Pour (2000) summarize the critical problems as follow: lack of a unified interconnection standard; lack of a proven economic model for selling components; and lack of a standard for component specification. Lots of problems remain in this field to be solved before reusable component software engineering can become more practical.

Traditional software development usually produces custom-made software that usually has the significant advantage of being optimally adapted to the user's business model, and it can fulfill all the requirements of the in-house proprietary knowledge or practices. But it also has the severe disadvantage of huge cost.

A software component is what is actually deployed, just like an isolated part of a system. It is different than an object that is almost never sold, bought, or deployed. A component could just as well use some totally different implementation technology, such as pure functions or assembly languages that follow some special flow, and there is nothing that looks like objects. So even though in some books and papers, the terms "component" and "object" are often used interchangeably, they are totally different. But when we talk about reuse, we have to deal with both of them together.

Modeling Transfer and Verification

Current modeling techniques and standards offer explicit definitions and notations to support software development, but few of them provide the capability to enable users to verify their works' consistency while drifting to other techniques that are needed in the next phases of the development; this leads to limited automation and inefficiency. Some researchers have dedicated their works to improving the situation.

Modeling and Understanding. The technique helps an engineer compare artifacts by summarizing where one artifact (such as a design) is consistent with and inconsistent with another artifact (such as source) (Murphy, Notkin, & Sullivan, 2001). Some other works develop the software reflection model technique to help engineers perform various software engineering tasks by exploiting, rather than removing the drift between design and implementation (Aho, Kernighan, & Weinberger, 1979; Murphy et al., 2001). Based on the similar concept, engineer might use a reverse engineering system to derive a high-level model from the source code (Wong, Tilley, Muëller, & Storet, 1995).

Automation. Only limited automation of the detail design and implementation is possible for now. A common problem of software development is how to derive executable software components from requirements, and how this process could be systematized. Under this consideration, formal methods might be the way to promise increasing reliability of software systems and the potential of automating the software development process (Jia, 1997). Koskimies, Systä, and Tuomi (1998) used an approach design by animation to construct new (or more complete) state diagrams along with the existing state diagrams and scenario diagrams.

Verification. Using Verification, (Cheng et al., 2000) diagrams with their specifications can be analyzed using automated techniques to validate behavior through simulation or to check for numerous properties of the diagrams, including inter- and intra-model consistency. Object Structure Model (Hayes & Coleman, 1991) is used to verify consistency between models in OMT methodology. In the Object Structure Model, the pre-condition and post-condition are defined and used (Deitel, Deitel, Nieto, Lin, & Sadhu, 2001). Another method is based on the Finite State Machine (FSM) (Atlee & Gannon, 1993). First, the user's requirements are expressed with Software Cost Reduction (SCR). Next, they are transformed into a Computational Tree Logic (CTL) machine. Finally, the validity of temporal logic is investigated (Deitel et al., 2001).

Software models are dynamically changed during the analysis/design, revision, and maintenance phases of the software lifecycle. Software tools at each phase of software development or maintenance employ their own formats to describe the software model information. Unfortunately, none of these standards is general enough to cover all phases of the software lifecycle, thus developers would have to adopt more than one standard for assistance to accomplish their work. We can also discover that un-successive gaps exist between these standards' applications. Figure 2 summarizes the relationship of the standards along with their positions during the lifecycle of software development. The notation ⊠ expresses that there is a need for modeling transfer between successive phases/models in a specific standard; the notation ⊗ points out the absence of consistency from one standard to the others.

New standards will surely keep emerging for new requirements of software engineering. It is clear that each of the modeling information expressed with a specific standard can only show part of the system information from its particular aspect. In this chapter, we would rather propose a unified system model to collaborate with various models from different well-accepted standards.

eXtensible Markup Modeling Language (XML)

XML (Deitel et al., 2001) is a standard language supported by the World Wide Web Consortium (W3C) with many useful features, such as application neutrality (vendor independence), user extensibility, ability to represent arbitrary and complex information, validation scheme for data structure, and human readability. Its main components are:

- Document Type Definition (DTD) specifies a set of tags and their constraints. Every document format is defined in DTD.
- XML Style Language (XSL) describes stylesheets for an XML document that is to take care of the document presentation. XSL is based on Document Style Semantics and Specification Language (DSSL) and Cascading Style Sheet (CSS).
- Xpointer and Xlink are defined anchors and links across XML documents.

Figure 2. The relationship of some standards and the lifecycle of software development.

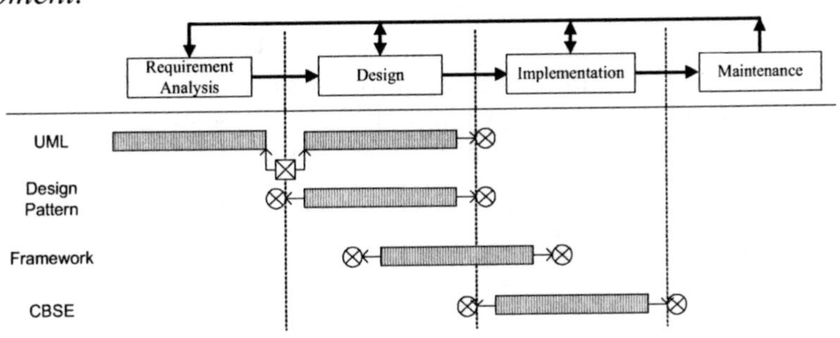

XML schema is a language that defines structure and constraints of the contents of XML documents. An XML schema consists of type definitions and element declarations. These can be used to assess the validity of a well-formed element and attribute information items, and furthermore may specify augmentations to those items and their descendants. Basically, the primary components of an XML schema are as follows:

- Simple type definitions, which cannot have element content and can not carry attributes.
- Complex type definitions, which allow elements as their content and may carry attributes.
- Attribute declarations, which are associations between a name and a simple type definition, together with occurrence information and optionally, a default value.
- Element declarations, which are associations of a name with a type definition, either simple or complex, an optional default value and a possibly empty set of identity-constraint definitions.

Suzuki and Yamamoto (1998) proposed an XML-based UML Exchange Format (UXF) to transform UML diagrams into XML. However, they focused on the format exchange between UML and XML representations, while our approach focuses on integration and unification of models.

THE XML-BASED UNIFIED MODEL

In our approach, an XML-based Unified Meta-Model (XUMM) is used to define the schema of the integration and unification of the modeling information from the adopted standard models, such as analysis and design models represented in UML, design pattern models, framework, etc., in each phase of the software life cycle. Each adopted well-known standard model may be represented by a set of submodels. (A submodel is called a model in this chapter.) For example, UML uses class diagram, object diagram, etc., as the submodels of analysis and design phases. Each model in adopted standards has its corresponding representation in XUM called a *view* of the unified model.

As shown in Figure 3, based on XUMM, these submodels are integrated, unified, and represented in an XML-based Unified Model (XUM). Various modeling information from adopted standards not only transformed into XUM but also unified. Semantics in each model should be described explicitly and transferred precisely in XUM.

The XUM should facilitate the following tasks:

1) the capturing of modeling information of models and transforming into views of XUM;

2) the two-way mapping of modeling information among models and XUM views;
3) the integration and unification of modeling information of different views in XUM;
4) the support of systematic manipulation;
5) the consistent checking of views represented in XUM; and
6) the reflection of changes of view information in XUM to models in each phase.

XML-Based Unified Meta-Model (XUMM)

Figure 4 shows the relationship of views in XUM. One of the merits of XUM is that it not only keeps the modeling information used in models (views) of each phase of software life cycle in a uniform format, but also the modeling information relationship across modes (views) are explicitly defined and represented.

The relationship of XUMM with an XUM is like DTD with XML document. XUMM defines the schema of XUM. Three major elements are defined to describe XUMM and XUM—*component, association,* and *unification relation*. Any object in XUM is called a component. The terms *component* and *association* describe the semantic information of components and their relationships, respectively. The *unification relation* is used to describe the relationship of different views.

In the XUMM, three *primitive* schemas are defined—*ComponentType, AssociationType,* and *Unification_linkType.* The ComponentType schema defines the necessary modeling semantic information and the types that are used to describe any object in our model. The XML schema of *ComponentType* is defined as follows:

```
<xsd:complexType name="ComponentType">
 <xsd:annotation>
```

Figure 3. The unification and integration of models into XUM.

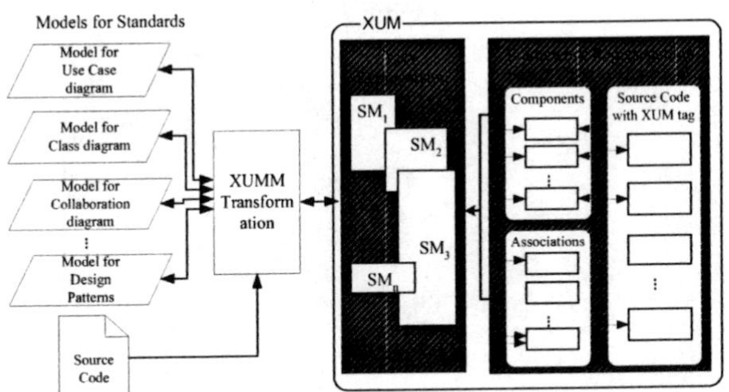

```
<xsd:documentation>Definition of primitive element: Component</
 xsd:documentation>
 </xsd:annotation>
 <xsd:attribute name="name" type="xsd:string" use="required"/>
 <xsd:attribute name="id" type="xsd:string" use="optional"/>
</xsd:complexType>
```

The *AssociationType* schema defines the needed information and the types that are used to describe the relationship of components. The XML schema of *AssociationType* is defined as follows:

```
<xsd:complexType name="AssociationType">
 <xsd:annotation>
 <xsd:documentation>Definition of primitive element: Association</
 xsd:documentation>
 </xsd:annotation>
 <xsd:attribute name="from" type="xsd:string"/>
 <xsd:attribute name="to" type="xsd:string"/>
</xsd:complexType>
```

Figure 4. The relationship among views in XUM.

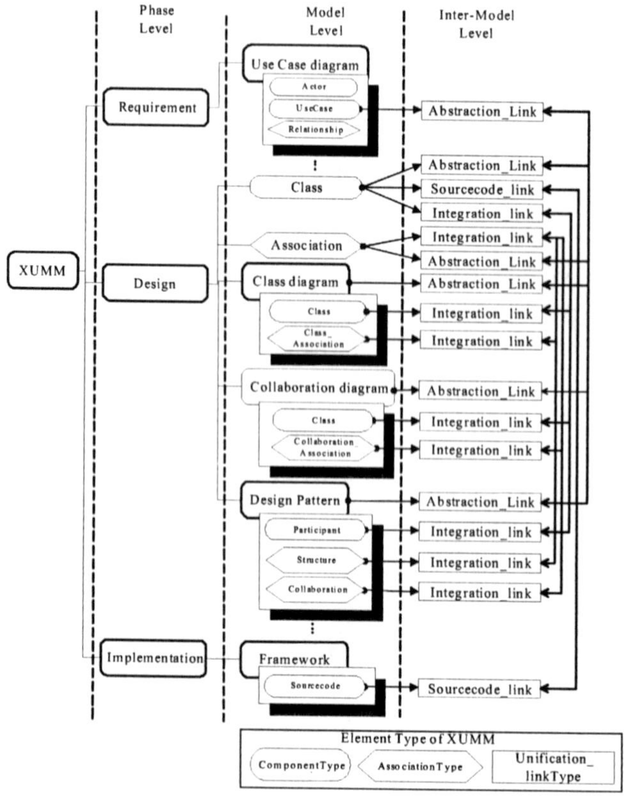

In order to show the relationship of the integration and unification of views in XUM, *Unification_linkType* is been defined. *Unification_linkType* schema defines the hyperlink relations between elements of XUM using a set of xlink. The *Unification_linkType* is defined as follows:

```
<xsd:complexType name="Unification_linkType">
 <xsd:annotation>
<xsd:documentation>Definition of primitive element: Unification_lin</
xsd:documentation>
 </xsd:annotation>
<xsd:complexType>
 <xsd:attribute name="xlink:type" type="(locator)" use="fixed" value=""/>
 <xsd:attribute name="xlink:arcole" type="xsd:CDATA" use="required"/>
 <xsd:attribute name="xlink:href" type="xsd:string" use="required"/>
 <xsd:attribute name="xlink:title" type="xsd:CDATA"/>
 <xsd:attribute name="xlink:from" type="xsd:NMTOKEN"/>
 <xsd:attribute name="xlink:to" type="xsd:NMTOKEN"/>
 </xsd:complexType>
</xsd:complexType>
```

Based on *Unification_linkType*, three types of links are defined: *Integration_link*, *Abstraction_link*, and *Sourcecode_link*. The *Integration_link* represents the link of a set of components or/and associations that have the same semantics but may be named or represented differently in different views. Its schema is defined as follows:

```
<xsd:element name="Integration_link" type="Unification_linkType"
maxOccurs="unbounded">
```

The *Abstraction_link* represents the link that connects component/association to a view that consists of a set of components and their association representing the details of that component at lower level of abstraction. Its schema is as follows:

```
<xsd:element name="Abstraction_link" type="Unification_linkType" minOccurs="0"
maxOccurs = "unbounded"/>
```

The *Sourcecode_ link* represents the link between the component and its corresponding source code and is defined as follows:

```
<xsd:element name="Sourcecode_link" type="Unification_linkType">
```

Based on integration, abstraction, and sourcecode links, the models adopted from various standards that might share some semantics but were not explicitly

Figure 5. The structure of XUMM.

```
<?xml version="1.0" encoding="UTF-8"?>
<xsd:schema xmlns:xsd="http://www.w3.org/2000/10/XMLSchema" xmlns:xlink="http://www.w3.org/
1999/xlink" elementFormDefault="qualified">
   <xsd:complexType name="ComponentType">
... ...
   <xsd:complexType name="AssociationType">
   ... ...
   <xsd:complexType name="Unification_linkType">
... ...   <xsd:element name="XUMM">... ...
        <xsd:element name="Requirement">
... ...          <xsd:element name="UseCase_Daigram">... ...
                    <xsd:element name="Actor" type="ComponentType" minOccurs="0"
maxOccurs="unbounded">
... ...
        <xsd:element name="Usecase" minOccurs="0" maxOccurs="unbounded">
... ...
<xsd:element name="Abstraction_link" type="Unification_linkType" minOccurs="0"
maxOccurs="unbounded"/>          <xsd:element name="Relationship" minOccurs="0"
maxOccurs="unbounded">
... ...
      <xsd:element name="Design">
... ...          <xsd:element name="Class" maxOccurs="unbounded">
... ...
            <xsd:element name="Attributes" minOccurs="0" maxOccurs="unbounded">
          ... ...
                    <xsd:element name="Integration_link" type="Unification_linkType"
maxOccurs="unbounded">
... ...
              <xsd:element name="Sourcecode_link" type="Unification_linkType">
... ...
              <xsd:element name="Abstraction_link" type="Unification_linkType" minOccurs="0"
maxOccurs="unbounded"/>
... ...          <xsd:element name="Association" maxOccurs="unbounded">... ...
        <xsd:element name="Class_Diagram" maxOccurs="unbounded">
... ...
          <xsd:element name="Collaboration_Diagram" minOccurs="0" maxOccurs="unbounded">
... ...
          <xsd:element name="Design_Pattern" minOccurs="0" maxOccurs="unbounded">
... ...
           <xsd:element name="Participation" maxOccurs="unbounded">
... ...
          <xsd:element name="Structure" maxOccurs="unbounded">
... ...
          <xsd:element name="Collaboration" maxOccurs="unbounded">
... ...
       <xsd:element name="Implementation">
... ...
       <xsd:element name="Framework" minOccurs="0" maxOccurs="unbounded">
... ...
        <xsd:element name="Sourcecode" maxOccurs="unbounded">
... ...
            <xsd:element name="Sourcecode_link" type="Unification_linkType">
... ...
</xsd:schema>
```

Table 1. Mapping of model elements and XUM elements.

Models/Standards	Model elements	XUM element Representations
UML Use Case diagram	Actor	<Actor>
	Use Case	<Usecase>
	Association	<Relationship type="association">
	Generalization	<Relationship type="generalization">
	Extend	<Relationship type="extend">
	Include	<Relationship type="include">
UML Class diagram, Collaboration diagram, Sequence diagram	Class Attribute Operation	<Class> <Attribute> <Operation>
	Interface	<Interface>
	Parameter	<Parameter>
	Association	<Class_Association type="association">
	Composition	<Class_Association type="composition">
	Generalization	<Class_Association type="generalization">
	Dependency	<Class_Association type="dependency">
	Message	<Message>
Design Pattern	Participants	<Participants>
	Structure	<Structure>
	Collaborations	<Collaboration>

represented can be integrated and unified in XUM. Therefore, when a model (view) gets changed, the changes can be reflected to other related models (views).

Each model adopted from standards has its corresponding XUM representation and its schema is defined in XUMM. Transforming modeling information into XUM is not a difficult task. Due to the space limitation, we only show the structure of XUMM in Figure 5 and its mapping rules in Table 1. The detailed schema of XUMM is shown in the Appendix.

XML-based Unified Model (XUM)

XML-based Unified Model (XUM) is the representation of artifacts of software systems defined in XUMM. These artifacts are the modeling information collected from models of standards used in each phase of the software life cycle. Firstly, each model is transformed into XUM and represented as a view in XUM. Since the semantics of models are explicitly captured and represented in views of XUM, XUM provides the capability of capturing modeling information of models. Secondly, the artifacts of views are integrated and unified into XUM.

Figure 6 shows the conceptual structure of XUM, which integrates and unifies models such as use case diagram, class diagram, collaboration diagram, design pattern, and source codes. The unification links provide two-way linking and the integration and unification of models.

Lastly, the manipulation of XUM is through XML's published interfaces based on the Document Object Model (DOM); i.e., DOM is the internal representation of XUM. The manipulation of semantic information of views of XUM is through the interfaces of DOM. Therefore, the systematic manipulation of XUM can be accomplished through the manipulation of DOM of XUM.

The unification link plays a very important role in tracking the related elements that reside in different views. These related elements may have abstraction relation, which are linked by abstraction_links. The views that share the same elements are linked by integration_links. The components in views that are related to source codes are linked by sourcecode_links. Based on these links, if any information in any view or any source code gets changed, the affected views can be reflected by following the links.

During software maintenance, modification to any model should reflect the related models, so the semantics in each model can be updated according to the modification. Therefore, the consistent checking of modeling information of views can be assisted. In addition, the impact analysis can be applied to the entire software system, including the impact on related source codes, the impact on related design information, and the impact on related requirement information.

AN EXAMPLE

In this section, to demonstrate the feasibility of our XUM approach, we have prepared the following example: the development and maintenance of a library subsystem – a book-loaning management software.

Figure 6. The conceptual structure of XUM.

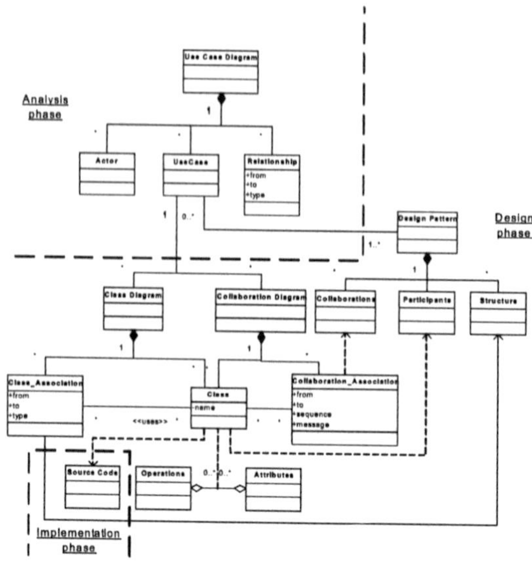

We assume the software development process has applied various popular standards in each phase, such as use case diagram, class diagram, collaboration diagram of UML, and design patterns. During the requirement analysis phase, a use case diagram is generated, as shown in Figure 7. Its corresponding XUM is represented in Figure 8. Note that, in Figure 8, the fields of abstraction_link are currently undefined and marked as "?," since we have not finished the integration and unification for views yet. These fields will be pointing to related views, such as design pattern view, collaboration diagram, and other views with abstraction relationships during integration and unification of views.

The class diagram, collaboration diagram, and design pattern diagram are created during the design phase. Figure 9 shows the class diagram, while Figures 10a and 10b respectively show the partial XUM representation of used classes and associations in Figure 9, and Figure 10c shows the XUM representation of class diagram.

Figure 7. A use case diagram of the system.

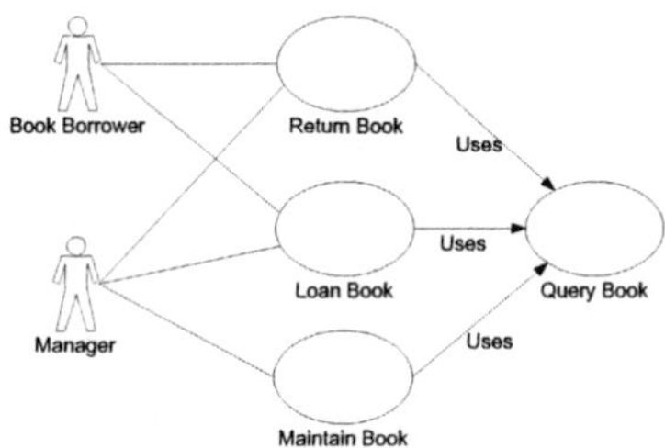

Figure 8. The XUM specification of the use case diagram.

```
<Requirement>
  <UseCase_Diagram>
     <Actor name="Book Borrower"/>
     <Actor name="Manager"/>
     <Usecase name="Loan Book">
  <Abstraction_link xlink:label="A_Loan_Book" xlink:title="Use Case of Loan_Book"
xlink:from="A_Loan_Book" xlink:to="? "/>
       <Abstraction_link xlink:from="A_Loan_Book" xlink:to=" ?"/>
... ...
    </Usecase>
     <Usecase name="Return Book">
... ...
     <Usecase name="Maintain Book">
... ...      <Usecase name="Query Book">
... ...
      <Relationship from="Book Borrower" to="Loan Book" type="association"/>
      <Relationship from="Book Borrower" to="Return Book" type="association"/>
      <Relationship from="Manager" to="Loan Book" type="association"/>
... ...
  </UseCase_Daigram>
 </Requirement>
```

Figure 9. The class diagram of the system.

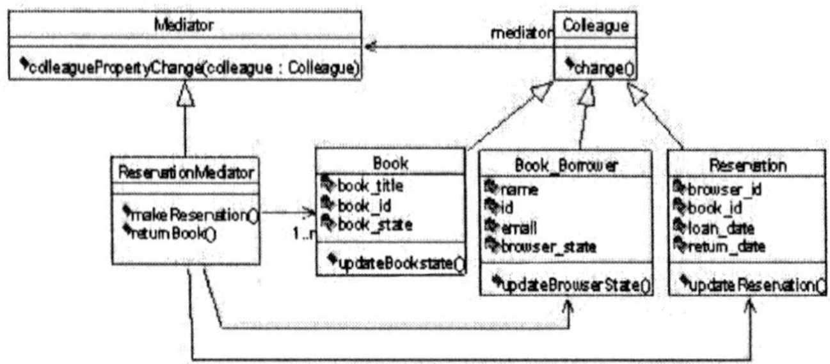

Figure 10a. XUM representation of classes.

```
<Class name="Mediator">
    <Integration_link xlink:label="D_Mediator" xlink:title="Class of Mediator"/>
    <Sourcecode_link xlink:from="D_Mediator" xlink:to="S_Mediator"/>
    <Operation name="colleaguePropertyChange(colleague:Colleague)" attribute="public" />
</Class>
<Class name="ReservationMediator">
    <Integration_link xlink:label="D_ReservationMediator" xlink:title="Class of ReservationMediator"/>
    <Sourcecode_link xlink:from="D_ReservationMediator" xlink:to="S_ReservationMediator"/>
    <Operation name="makeReservation()" attribute="public" />
    <Operation name="returnBook()" attribute="public" />
</Class>
<Class name="Book">
    <Integration_link xlink:label="D_Book" xlink:title="Class of Book"/>
    <Sourcecode_link xlink:from="D_Book" xlink:to="S_Book"/>
    <Attributes name="book_id" type="String" attribute="private"/>
    <Attributes name="book_title" type="String" attribute="private"/>
    <Attributes name="book_state" type="Boolean" attribute="private"/>
    <Operations name="updateBookState()" attribute="public" />
</Class>
<Class name="Book_Borrower">
    <Integration_link xlink:label="D_Book_Borrower" xlink:title="Class of Book_Borrower "/>
    <Sourcecode_link xlink:from="D_Book_Borrower" xlink:to="S_Book_Borrower "/>
    <Attributes name="name" type="String" attribute="private"/>
    <Attributes name="id" type="String" attribute="private"/>
    <Attributes name="E-mail" type="String" attribute="private"/>
    <Attributes name="browser_state" type="Boolean" attribute="private"/>
    <Operations name="updateBrowserState()" attribute="public" />
</Class>
<Class name="Reservation">
    <Integration_link xlink:label="D_Reservation" xlink:title="Class of Reservation "/>
    <Sourcecode_link xlink:from="D_Reservation" xlink:to="S_ Reservation "/>
    <Attributes name="Browser_id" type="String" attribute="private"/>
    <Attributes name="Book_id" type="String" attribute="private"/>
    <Attributes name="Loan_data" type="Data" attribute="private"/>
    <Attributes name="Return_dead_line" type="Date" attribute="private"/>
    <Operations name="updateReservation()" attribute="public" />
</Class>
<Class name="Colleague">
    <Integration_link xlink:label="D_Colleague" xlink:title="Class of Colleague"/>
    <Sourcecode_link xlink:from="D_Colleague" xlink:to="S_Colleague "/>
    <Operations name="change()" attribute="public" />
</Class>
```

The design pattern *Mediator* has been applied to this design, which happens to cover the same set of classes shown in Figure 9 in this example. The XUM representation of Mediator is shown in Figure 11. Figure 12 shows the collaboration diagram of the system, and its XUM representation is shown in Figure 13.

The capturing of modeling information from models and transforming them into XUM is quite systematic and straightforward. From previous diagrams and their corresponding XUM representations, each model adopted from standards has its corresponding XUM view. The view in XUM has explicitly and honestly represented the semantics of components and their relations, which may be implicitly represented in the adopted standard model. The naming of elements in models and XUM views is the same. Therefore, the two-way mapping between models and views has been constructed in our XUM approach.

Figure 10b. XUM representation of associations.

```
<Association from=“Mediator” to=“ReservationMediator”>
  <Intgration_link xlink:label=“Mediator_ReservationMediator” xlink:title“Association:
Mediator_ReservationMediator”/>
</Association>
<Association from=“ReservationMediator” to=“Reservation”>
  <Intgration_link xlink:label=“ReservationMediator_Reservation” xlink:title=“Association:
ReservationMediator_Reservation”/>
</Association>
<Association from=“ReservationMediator” to=“Book”>
  <Intgration_link xlink:label=“ReservationMediator_Book” xlink:title=“Association:
ReservationMediator_Book”/>
</Association>
<Association from=“ReservationMediator” to=“Book_Borrower”>
  <Intgration_link xlink:label=“ReservationMediator_Book_Borrower”
xlink:title=“Association: ReservationMediator_Book_Borrower”/>
</Association>
<Association from=”Reservation” to=“Colleague”>
  <Intgration_link xlink:label=“Reservation_Colleague” xlink:title=“Association:
Reservation_Colleague”/>
</Association>
<Association from=“Book” to=“Colleague”>
<Intgration_link xlink:label=“Book_Colleague” xlink:title=“Association: Book_Colleague”/>
</Association>
<Association from=“Book_Borrower” to=“Colleague>
  <Intgration_link xlink:label=“Book_Borrower_Colleague” xlink:title=“Association:
Book_Borrower_Colleague”/>
</Association>
<Association from=“Colleague” to=“Mediator”>
  <Intgration_link xlink:label=“Colleague_Mediator” xlink:title=“Association:
Colleague_Mediator”/>
</Association>
```

As shown in the XUM representation in previous figures, the unification links (such as the tags <Integration_link>, <Abstraction_link>, and <Sourcecode_link>) are used to link the components that share some semantic information. For example, class *ReservationMediator* in class diagram view has links from other views that have used it, such as *Mediator* design pattern view, collaboration diagram view, etc. Similarly, it has a link to source code segment that represents the class.

When modeling information is not complete, some of the tag <Unification_link> may be undefined. However, these undefined links are very valuable indications to

Figure 10c. The XUM specification of class diagram.

```
   <Class_Diagram>
    <Class name="Mediator">
<Integration_link xlink:href="D_Mediator"/>
    <Class name="ReservationMediator">
… …
   <Class name="Book">
… …
   <Class name="Book_Borrower">
… …
   <Class name="Reservation">
… …
   <Class name="Colleague">
… …
    <Class_Association from="Mediator" to="Reservation Mediator" type="generalization" client="1">
         <Integration_link xlink:title="Mediator_ ReservationMediator" xlink:lable=" Association of Mediator_ ReservationMediator" xlink:href="Mediator_ReservationMediator" xlink:from="D_Mediator" xlink:to="D_ ReservationMediator" />
    <Class_Association from="ReservationMediator" to="Reservation" type="dependency" client="0..n">
… …
    <Class_Association from="ReservationMediator" to="Book" type="dependency" client="1" >
… …
    <Class_Association from="ReservationMediator" to="Book_Borrower" type="dependency" client="1" >
… …
  <Class_Association from="Book" to="Colleague" type="generalization" client="1" >
… …
  <Class_Association from="Book_Borrower" to="Colleague" type="generalization" client="1">
… …
  <Class_Association from="Reservation" to="Colleague" type="generalization" client="1" >
… …
  <Class_Association from="Colleague" to="Mediator" type="dependency" client="1">
… …
  </Class_Diagram>
```

Figure 11. The XUM representation of design pattern Mediator.

```
<Design_Pattern name="Mediator" dominator="Loan_Book/Return_Book">
    <Abstraction_link xlink:title="Design_Pattern_Loan_Book" xlink:title="Design Pattern of
Loan_Book" xlink:href="A_Loan_Book"/>
    <Abstraction_link xlink:title="Design_Pattern_Return_Book" xlink:title="Design Pattern of
Return_Book" xlink:href="A_Return_Book"/>
  <Participant name="Mediator" role="Mediator">
    <Integration_link xlink:href="D_Mediator"/>
</ Participant >
    <Participant name="ReservationMediator" role="ConcreteMediator">
… …
  <Participant name="Colleague" role="Concrete classes">
… …
  <Participant name="Reservation" role="Colleague classes">
… …
  <Participant name="Book" role="Colleague classes">
… …
  <Participant name="Book_Borrower" role="Colleague classes">
… …
    <Structure from="Mediator" to="ReservationMediator" type="generalization" client="1">
      <Integration_link xlink:href="Mediator_ReservationMediator"/>
</Structure>
    <Structure from="ReservationMediator" to="Reservation" type="dependency" client="0..n">
… …
    <Structure from="ReservationMediator" to="Book" type="dependency" client="1">
… …
    <Structure from="ReservationMediator" to="Book_Borrower" type="dependency" client="1">
… …
<Structure from="Reservation" to="Colleague" type="generalization" client="1">
… …
<Structure from="Book" to="Colleague" type="generalization" client="1">
… …
<Structure from="Book_Borrower" to="Colleague" type="generalization" client="1">
… …
<Structure from= "Colleague" to="ReservationMediator" type="dependency" client="1">
… …
    <Collaboration from="ReservationMediator" to="Reservation" sequence="1"
message="returnBook()"/>
      <Integration_link xlink:href="ReservationMediator_Reservation"/>
  </Collaboration>
    <Collaboration from="ReservationMediator" to="Book" sequence="2"
message="updateReservation()">
… …
    <Collaboration from="ReservationMediator" to="Book_Borrower" …>
… …
  </Design_Pattern>
…
<Implementation>
<Framework>
<Source_code name=" ReservationMediator ">
 <Sourcecode_link xlink:label="S_ReservationMediator " xlink:title="Source code of
ReservationMediator" xlink:type="arc" xlink:from="S_ReservationMediator "
xlink:to="D_ReservationMediator " />public class ReservationMediator { …}</Source_code>
…
```

software engineers that the system is in incomplete situation, so some enhancement can be made.

Applying XUM to Software Maintenance

Software maintenance is an inevitable process and considered as an iteration of the software process. Software maintenance and evolution involve the identification or discovery of program requirements and/or design specifications that can aid in understanding, modifying, and adding features to the old programs. In addition, the factors that affect the software development equally affect the software maintenance. In particular, when the maintenance has been implemented on large

Figure 12. The collaboration diagram of the system.

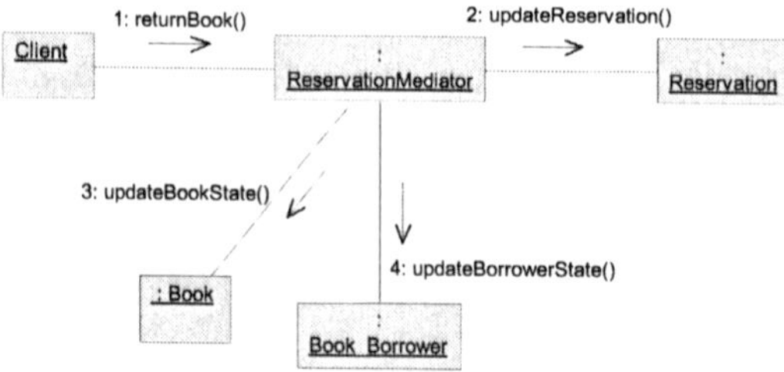

Figure 13. The XUM specification of collaboration diagram.

```
<Collaboration_Diagram>
<Class name="Mediator">
<Integration_link xlink:href="D_Mediator"/>
</Class>
<Class name="ReservationMediator">
… …
<Class name="Book">
… …
<Class name="Book_Borrower">
… …
<Class name="Reservation">
… …
<Class name="Colleague">
… …
<Collaboration_Association from="ReservationMediator" to="Reservation" sequence="1"
message="returnBook()"/>
<Integration_link xlink:href="ReservationMediator_Reservation"/>
</Collaboration_Association ><Collaboration_Association from="ReservationMediator"
to="Book" sequence="2" message="updateReservation()">
… …
<Collaboration_Association from="ReservationMediator" to="Book_Borrower" ...>
… …
</Collaboration_Diagram>
```

scale systems, the maintenance problems may become a disaster and unmanageable. Therefore, the costs of software maintenance are usually much greater than developing new similar software (Sommerville, 1996). There are some reasons for this:

- Maintenance staffs are inexperienced and unfamiliar with this software.
- Suitable tools are lacking.
- The software structure is difficult to understand.
- The related documentations are usually unreliable and inconsistent, which offer very limited assistance to software maintenance.

One of the difficulties in software maintenance is the maintenance of consistency of various documents, including requirement documents, design documents, comments in source codes, and source codes. However, these documents are usually not existing or inconsistent; source codes are usually the only reliable source that provides the most updated information. Without a mechanism to enforce the existence and consistency of these documents, the software maintenance problem has no way of being solved.

Here we use the same example to show how XUM can help on software maintenance. We assume a new function that can notify users who have reserved a book when it is returned to library. The component *Notification* is added to the collaboration diagram shown in Figure 12 and Figure 14. The resulting XUM of modified collaboration diagram is shown in Figure 15. A new <Class>, Notification, is specified. However, its <Integration_link>, which is supposed to point to class specification of XUM, is undefined at the current stage since its class specification is not specified yet. A new <Collaboration_Association> is defined for *ReservationMediator* and *Notification*.

From the missing information of <Integration_link> to class specification, the change can be reflected to the class diagram view shown in Figure 10a and 10b. Figure 10a does not have the specification of class *Notification*. Figure 10b is missing <Class> and <Class_Association> for class *Notification* also. In order to

Figure 14. The collaboration diagram after modification.

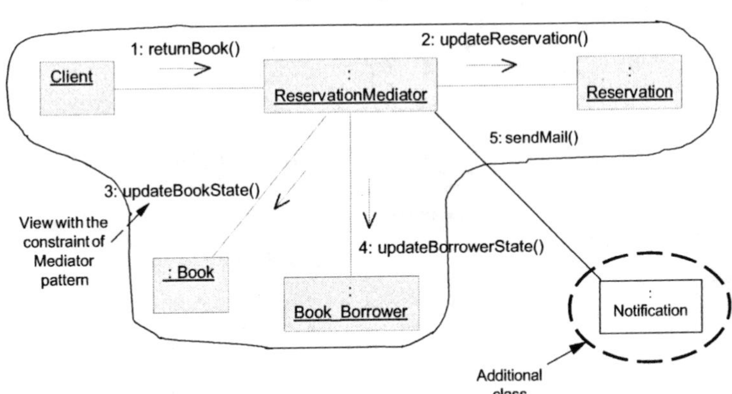

be consistent with the collaboration diagram view, software maintainers need to modify the class diagram. Figure 16 shows the modified class diagram. Figures 17a and 17b show the modified class and class diagram XUM representation.

Due to the <Class_Association> in the class diagram and <Collaboration_Association> in the collaboration diagram, that which is related to class *Notification* is related to class *ReservationMediator* also. Its <Association> has an integration link to <Collaboration> and <Structure> in design pattern that is related to class *ReservationMediator*. The <Collaboration> and <Struc-

Figure 15. The XUM representation of collaboration diagram after modification.

```
<Collaboration_Diagram dominator="Return_Book">
    <Abstraction_link xlink:label="Collaboration:Return_Book" xlink:title="Collaboration Diagram of
        Return_Book" xlink:from="Collaboration:Return_Book" xlink:to="A_Return_Book"/>
        ... ...
    <Class name="Colleague">
        <Integration_link xlink:href="D_Colleague" />
    </Class>
    <Class name="Notification">
        <Integration_link xlink:href="?" />
    </Class>
        ... ...
    <Collaboration_Association from="ReservationMediator" to="Book_Borrower" sequence="4"
            message="updateBookState()">
        <Integration_link xlink:from="D_ReservationMediator " xlink:to="D_Book_Borrower" />
    </Collaboration_Association>
    <Collaboration_Association from="ReservationMediator" to="Notification" sequence="5"
            message="sendMail()">
        <Integration_link xlink:from="D_ReservationMediator " xlink:to="?" />
    </Collaboration_Association>
</Collaboration_Diagram>
```

Figure 16. The class diagram after modification.

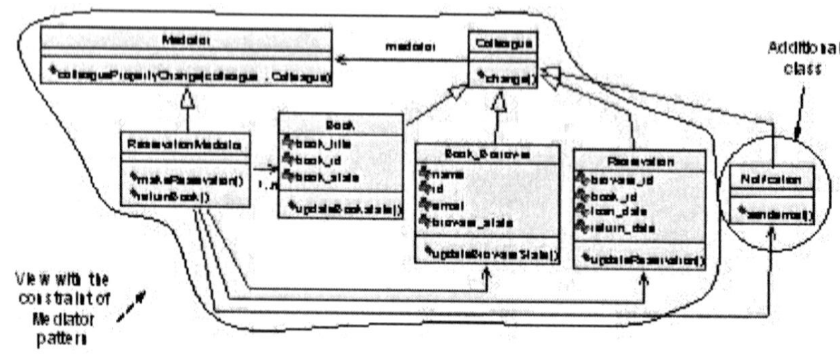

ture> from *ReservationMediator* to *Notification* are missing. From the change of <Collaboration> and <Structure>, the corresponding <Participant> needs to be reflected in the change. The result of design pattern view for Mediator is shown in Figure 18.

Since the example of maintenance, XUM through unifying the modeling information and building related unification_link. If a designer makes any modification, the change will be reflected to corresponding.

Figure 17a. The XUM specification of class after modification.

```
... ...
<Class name="Colleague">
    <Integration_link xlink:label="D_Colleague" xlink:title="Class of Colleague"/>
    <Sourcecode_link xlink:from="D_Colleague" xlink:to="S_Colleague "/>
    <Operations name="change()" attribute="public" />
</Class>
<Class name="Notification ">
    <Integration_link xlink:label="D_Notification" xlink:title="Class of Notification"/>
    <Sourcecode_link xlink:from="D_Notification" xlink:to="S_Notification"/>
    <Operations name="sendMail()" attribute="public" />
</Class>
```

Figure 17b. The XUM specification of class diagram after modification.

```
<Class_Diagram dominator="Return_Book">
    <Abstraction_link xlink:label="CD_Return_Book" xlink:title="Class Diagram of Return_Book"
          xlink:from="CD_Return_Book" xlink:to="A_Return_Book"/>
    ... ...
    <Class name="Colleague">
        <Integration_link xlink:href="D_Colleague" />
    </Class>
    <Class name="Notification">
        <Integration_link xlink:href="D_Notification" />
    </Class>
    ... ...
    <Class_Association from="Colleague" to="Mediator" type="dependency" client="1">
        <Integration_link xlink:href="Colleague_Mediator" xlink:from="D_Colleague" xlink:to="D_Mediator"
            />
    </Class_Association>
    <Class_Association from="ReservationMediator" to="Notification" type="dependency" client="1">
        <Integration_link xlink:href="ReservationMediator_Notification" xlink:from=
            "D_ReservationMediator" xlink:to="D_Notification" />
    </Class_Association>
    <Class_Association from="Notification" to="Colleague" type="generalization" client="1">
        <Integration_link xlink:href="Notification_Colleague" xlink:from="D_Notification"
            xlink:to="D_Colleague" />
    </Class_Association>
</Class_Diagram>
```

FUTURE TRENDS

According to the discussion and validation above, we believe that the XUM approach proposed in this chapter can be extended and benefit more activities in various processes for software development, especially software maintenance. We suggest the future directions based on our unified model and list them in the following:

1. *Linking reuse technologies into the software development process with unified model*
   Applying reuse at an earlier phase of the software life cycle can reduce the software cost and increase the software productivity greatly. Without integrating and unifying the models used in each phase, the links from the requirement, design, to implementation are missing. The reuse of models at the early phase has no way to link to its corresponding source codes. With the support of the unified model, the integration with software reuse technologies can be another direction that needs to be studied further.
2. *The modeling of design patterns*
   Most of the available design patterns are not yet formally specified and modeled. Although we have tried to capture the information of the design patterns in our model, they are not well specified yet. In order to accomplish

Figure 18. The design pattern view of XUM after modification.

```
    ... ...
  <Design_Pattern name="Mediator" dominator="_System">
      <Abstraction_link xlink:title="Design_Pattern_System" xlink:title="Design Pattern of _System "
xlink:href="..."/>
      <Participant name="Notification" role="Concrete class"/>
         <Integration_link xlink:href="D_Notification"/>
      </ Participant >
         ... ...
      <Structure from="Mediator" to="ReservationMediator" type="generalization" client="1">
           <Integration_link xlink:href="Mediator_ReservationMediator"/>
      </Structure>
         ... ...
        <Structure from="ReservationMediator" to="Notification" type="dependency" client_cardinality="1">
           <Integration_link xlink:href="ReservationMediator_Notification"/>
      </Structure>
      <Collaboration from="Client" to="Mediator" sequence="1" message="returnBook()"/>
           <Integration_link xlink:href="_Mediator"/>
      </Collaboration>
         ... ...
        <Collaboration from="ReservationMediator" to="Notification" sequence="5" message= "sendMail()">
            <Integration_link xlink:href="ReservationMediator_ Notification"/>
        </Collaboration>
    ... ...
</Pattern>
... ...
```

the further analysis on design patterns while applying them to software development or maintenance, the modeling of design patterns is another important issue. In particular, when it is applied to other standards, there are some extra issues related to integration and unification.

3. *The verification of consistency of models*
   Making applied models in the software process consistent is a very important task. Some researchers have worked in this direction, such as Holzmann (1997), who offered a verification system for models of distributed software systems. Gunter, Gunter, Jackson, and Zave (2000) presented a reference model for requirements and specifications. Our unified model helps the consistency checking among views, but offers very limited assistance to verification. The verification of consistency of models is one of the future directions of verification of consistency of models.
4. *XML-based program analysis*
   Most current source codes are kept in file format with no structure representation. Therefore, not only requirement documents and design documents but also the comments written in source codes have no explicit link to source code segments. Representing programs in XML with a unified model that links source codes to related models/documents can facilitate the program analysis. Since XML documents are represented in DOM and the compilers of XML exist, efforts to implement analysis tool sets are much easier.
5. *The XUM-based software environment*
   Providing a software environment that supports our XUM approach is another important future.

CONCLUSIONS

In this chapter, we have proposed an XML-based unified model that can integrate and unify a set of well-accepted standards into a unified model represented in a standard and well-accepted language, XML. The survey of these adopted standards and their roles in the software life cycle are presented in this chapter as well.

The XUM can facilitate the following tasks:

1) The capturing of modeling information of models and transforming into views of XUM.
2) The two-way mapping of modeling information among models and XUM views.
3) The integration and unification of modeling information of different views in XUM.
4) The support of systematic manipulation.
5) The consistent checking of views represented in XUM.

6) The reflection of changes of view information in XUM to models in each phase.

With the above capability, XUM can improve software maintainability. The linking among models in each phase in XUM can help reuse at earlier stages of software development.

REFERENCES

Aho, A.V., Kernighan, B.W., & Weinberger, P.J. (1979). Awk—A pattern scanning and processing language. *Software-Practice and Experience*, *9*(4), 267-280.

Atlee, J.M., & Gannon, J. (1993). State-based model checking of event-driven system requirements. *IEEE Transactions on Software Engineering*, *19*(1), 24-40.

Booch, G. (1991). *Object-oriented design with applications*. Redwood City, CA: Benjamin/Cummings.

Booch, G. (1994). *Object-oriented analysis and design with applications,* 2nd ed. Redwood City, CA: Benjamin/Cummings.

Bourdeau, R.H., & Cheng, B.H.C. (1995). A formal semantics for object model diagrams. *IEEE Transactions on Software Engineering*, *21*(10), 799-821.

Chen, D.J., & Chen, T. K. (1994, May). An experimental study of using reusable software design frameworks to achieve software reuse. *Journal of Object-Oriented Programming*, *7*(2), 56-67.

Cheng, B.H.C., Campbell, L.A., & Wang, E.Y. (2000). Enabling automated analysis through the formalization of object-oriented modeling diagrams. In the *Proceedings International Conference on Dependable Systems and Networks 2000 (DSN 2000)*, IEEE, 305-314.

Chu, W.C., Lu, C.W., Chang, C.H., & Chung, Y.C. (2001). Pattern based software re-engineering. *Handbook of Software Engineering and Knowledge Engineering*, Vol. 1. Skokie, IL: Knowledge Systems Institute.

Chu, W.C., Lu, C.W, Yang, H., & He, X. (2000). A formal approach to component retrieval and integration. *Journal of Software Maintenance*, *12*(6), 325-342.

Connolly, D. (2001). The extensible markup language (XML). The World Wide Web Consortium. Retrieved August 21, 2001 from http://www.w3.org/XML.

Deitel, H., Deitel, P., Nieto, T., Lin, T., & Sadhu, P. (2001). *XML How To Program*. Upper Saddle River, NJ : Prentice Hall.

Do-Hyoung, K., & Kiwon, C. (1996). A method of checking errors and consistency in the process of object-oriented analysis. In *Proceedings of the 3rd Asia-Pacific Software Engineering Conference (APSEC '96)*, IEEE, 208-216.

Gamma, E., Helm, R., Johnson, R., & Vlissides, J. (1995). *Design Patterns: Elements of Reusable Object-Oriented Software*. Reading, MA: Addison-Wesley.

Gunter, C.A., Gunter, E.L., Jackson, M., & Zave, P. (2000, May/June). A reference model for requirements and specifications. *IEEE Software, 17*(3), 37-43.

Hartrum, T.C., & Bailor, P. D. (1994). Teaching formal extensions of informal-based object-oriented analysis methodologies. In the *Proceedings of Computer Science Education*, 389-409.

Hayes, F., & Coleman, D. (1991). Coherent models for object-oriented analysis. In *Proceedings on ACM OOPSLA '91*. ACM, 171-183.

Holland, I.M. (1993). *The design and representation of object-oriented components*. PhD thesis, Northeastern University. Retrieved March 20, 1996 from http://www.ccs.neu.edu/home/lieber/theses-index.html.

Holzmann, G.J. (1997). The model checker SPIN. *IEEE Transactions on Software Engineering, 23*(5), 279-295.

Jia, X. (1997). A pragmatic approach to formalizing object-oriented modeling and development. In *Proceedings of the COMPSAC '97—21st International Computer Software and Applications Conference*. IEEE, 240-245.

Jia, X. (1998a). *A Tutorial of ZANS—A Z Animation System*. Retrieved Feb. 25, 1998 from http://venus.cs.depaul.edu/fm/zans.html.

Jia, X. (1998b). *ZTC: A Type Checker for Z Notation, User's Guide,* Version 2.03. Retrieved August 12, 1998 from http://venus.cs.depaul.edu/fm/ztc.html.

Jia, X., & Skevoulis, S. (1998). *VENUS: A Code Generation Tool, User Guide,* Version 0.1. Retrieved August 25, 1998 from http://venus.cs.depaul.edu/fm/venus.html.

Johnson, R.E., & Foote, B. (1988). Designing reusable class. *Journal of Object-Oriented Programming, 1*(2), 22-35.

Koskimies, K., Systä, T., & Tuomi, J. (1998). Automated support for modeling oo software. *IEEE Software, 15*(1), 87- 94.

Lano, K., & Malik, N. (1997). Reengineering legacy applications using design patterns. In the *Proceedings of the 8th International Workshop on Software Technology and Engineering Practice*, IEEE, 326-338.

Lear, A.C. (1999). XML seen as integral to application integration. *IT Professional, 2*(5), 12-16.

Meyer, B. (1990). Tools for the new culture: Lessons from the design the Eiffel libraries. *Communications of the ACM, 33*(9), 68-88.

Moreira, A.M.D., & Clark, R.G. (1996). Adding rigour to object-oriented analysis. *Software Engineering Journal, 11*(5), 270-280.

Moser, S., & Nierstrasz, O. (1996). The effect of object-oriented frameworks on developer productivity. *IEEE Computer, 29*(9), 45-51.

Murphy, G.C., Notkin, D., & Sullivan, K.J. (2001). Software reflexion models: Bridging the gap between design and implementation. *IEEE Transactions on Software Engineering, 27*(4), 364-380.

OMG, Object Management Group. (2001). *OMG Unified Modeling Language Specification,* Version 1.4. Retrieved July 16, 2001 from http://www.omg.org/technology/documents/recent/omg_modeling.htm.

Ossher, H., Kaplan, M., Harrison, W., Katz, A., & V. Kruskal. (1995). Subject-oriented composition rules. In the *Proceedings of Object-Oriented Programming Systems, Languages and Applications Conference, special issue of SIGPLAN Notices.* ACM, 235-250.

Paul, S., & Prakash, A. (1994, June). Framework for source code search using program patterns. *IEEE Transactions on Software Engineering, 22*(6), 463-475.

Rine, D. C. (1997). Supporting reuse with object technology. *IEEE Computer, 30*(10), 43-45.

Shroff, M., & France, R. B. (1997, August). Towards a formalization ofUML class structures in Z. In *Proceedings Twenty-First Annual International Computer Software and Applications Conference (COMPSAC'97).* IEEE, 646-651.

Smith, R., Szyperski, C., Meyer, B., & Pour, G. (2000). Component-based development? refining the blueprint. In *Proceedings ofTechnology of Object-Oriented Languages and Systems (TOOLS 34'00).* IEEE, 563-566.

Sommerville, I. (1996). *Software Engineering*, 5th Edition. Reading, MA: Addison-Wesley.

Spivey, J.M. (1992). *The Z notation*, 2nd Edition. Upper Saddle River, NJ: Prentice Hall.

Suzuki, J., & Yamamoto, Y. (1998). Making UML models exchangeable over the internet with XML: UXF approach. In the *Proceedings on UML'98: Beyond the Notation—International Workshop*, 65-74.

Szyperski, C., & Pfister, C. (1997). Workshop on component-oriented programming, Summary In M. Muhlhauser (Ed.), *Special Issues in Object-Oriented Programming–ECOOP96 Workshop Reader*. Heidelberg, Germany: dpunkt-Verlag.

Wang, E.Y., & Cheng, B.H.C. (1998). A rigorous object-oriented design process. In the *Proceedings of International Conference on Software Process,* Naperville. Retrieved from http://www.cse.msu.edu/~chengb/pubs.html.

Wirfs-Brock, R. J., & Johnson, R. E. (1990). Surveying current research in object-oriented design. *Communications of the ACM, 33*(9), 105-124.

Wong, K., Tilley, S.R., MuÈller, H.A., & Storey, M.D. (1995). Structural redocumentation: A case study. *IEEE Software, 12*(1), 46-54.

Xiao, C. (1994). *Adaptive software: Automatic navigation through partially specified data structures*. PhD thesis, Northeastern University. Retrieved May 7, 2001, from http://www.ccs.neu.edu/home/lorenz/center/ index.html.

Yau, S.S., & Dong, N. (2000). Integration in component-based software development using design patterns. In the *Proceedings on the 24th Annual International Computer Software and Applications Conference (COMPSAC 2000)*, 369-374.

APPENDIX
XUMM SCHEMA

```
<?xml version="1.0" encoding="UTF-8"?>
<xsd:schema xmlns:xsd="http://www.w3.org/2000/10/XMLSchema" xmlns:xlink="http://
www.w3.org/1999/xlink" elementFormDefault="qualified">
  <xsd:complexType name="ComponentType">
    <xsd:annotation>
      <xsd:documentation>Definition of primitive element: Component</xsd:documentation>
    </xsd:annotation>
    <xsd:attribute name="name" type="xsd:string" use="required"/>
    <xsd:attribute name="id" type="xsd:string" use="optional"/>
  </xsd:complexType>
  <xsd:complexType name="AssociationType">
    <xsd:annotation>
      <xsd:documentation>Definition of primitive element: Association</xsd:documentation>
    </xsd:annotation>
    <xsd:attribute name="from" type="xsd:string"/>
    <xsd:attribute name="to" type="xsd:string"/>
  </xsd:complexType>
  <xsd:complexType name="Unification_linkType">
    <xsd:annotation>
      <xsd:documentation>Definition of primitive element: Unification_link</
xsd:documentation>
    </xsd:annotation>
  <xsd:complexType>
      <xsd:attribute name="xlink:type" type="(locator)" use="fixed" value=""/>
      <xsd:attribute name="xlink:arcole" type="xsd:CDATA" use="required"/>
    <xsd:attribute name="xlink:href" type="xsd:string" use="required"/>
      <xsd:attribute name="xlink:title" type="xsd:CDATA"/>
      <xsd:attribute name="xlink:from" type="xsd:NMTOKEN"/>
      <xsd:attribute name="xlink:to" type="xsd:NMTOKEN"/>
    </xsd:complexType>
</xsd:complexType >
  <xsd:element name="XUMM">
    <xsd:complexType>
      <xsd:sequence>
        <xsd:element name="Requirement">
          <xsd:complexType>
            <xsd:sequence>
              <xsd:element name="UseCase_Daigram">
                <xsd:complexType>
                  <xsd:sequence>
                    <xsd:element name="Actor" type="ComponentType" minOccurs="0"
maxOccurs="unbounded">
                      <xsd:annotation>
```

```
                    <xsd:documentation>Element(component) of XUM for requirement:
UML Use Case Daigram: Actor</xsd:documentation>
                  </xsd:annotation>
                </xsd:element>
                <xsd:element name="Usecase" minOccurs="0" maxOccurs="unbounded">
                  <xsd:annotation>
                    <xsd:documentation>Element(component) of XUM for requirement:
UML Use Case Daigram: Usecase</xsd:documentation>
                  </xsd:annotation>
                  <xsd:complexType>
                    <xsd:complexContent>
                      <xsd:extension base="ComponentType">
                        <xsd:sequence>
                          <xsd:element name="Abstraction_link"
type="Unification_linkType" minOccurs="0" maxOccurs="unbounded"/>
                        </xsd:sequence>
                      </xsd:extension>
                    </xsd:complexContent>
                  </xsd:complexType>
                </xsd:element>
                <xsd:element name="Relationship" minOccurs="0"
maxOccurs="unbounded">
                  <xsd:annotation>
                    <xsd:documentation>Element(association) of XUM for requirement:
UML Use Case Diagram association</xsd:documentation>
                  </xsd:annotation>
                  <xsd:complexType>
                    <xsd:complexContent>
                      <xsd:extension base="AssociationType">
                        <xsd:attribute name="type">
                          <xsd:simpleType>
                            <xsd:restriction base="xsd:NMTOKEN">
                              <xsd:enumeration value="association"/>
                              <xsd:enumeration value="extend"/>
                              <xsd:enumeration value="generalization"/>
                              <xsd:enumeration value="include"/>
                            </xsd:restriction>
                          </xsd:simpleType>
                        </xsd:attribute>
                      </xsd:extension>
                    </xsd:complexContent>
                  </xsd:complexType>
                </xsd:element>
              </xsd:sequence>
            </xsd:complexType>
          </xsd:element>
        </xsd:sequence>
      </xsd:complexType>
    </xsd:element>
    <xsd:element name="Design">
      <xsd:complexType>
        <xsd:sequence>
          <xsd:element name="Class" maxOccurs="unbounded">
```

```
<xsd:annotation>
  <xsd:documentation>Element(component) of XUM for design: Class</
xsd:documentation>
</xsd:annotation>
<xsd:complexType>
  <xsd:complexContent>
    <xsd:extension base="ComponentType">
      <xsd:sequence>
        <xsd:element name="Attributes" minOccurs="0"
maxOccurs="unbounded">
          <xsd:complexType>
            <xsd:complexContent>
              <xsd:extension base="ComponentType">
                <xsd:attribute name="type" type="xsd:string"/>
                <xsd:attribute name="limit">
                  <xsd:simpleType>
                    <xsd:restriction base="xsd:NMTOKEN">
                      <xsd:enumeration value="public"/>
                      <xsd:enumeration value="protect"/>
                      <xsd:enumeration value="private"/>
                    </xsd:restriction>
                  </xsd:simpleType>
                </xsd:attribute>
              </xsd:extension>
            </xsd:complexContent>
          </xsd:complexType>
        </xsd:element>
        <xsd:element name="Operations" minOccurs="0"
maxOccurs="unbounded">
          <xsd:complexType>
            <xsd:complexContent>
              <xsd:extension base="ComponentType">
                <xsd:attribute name="limit">
                  <xsd:simpleType>
                    <xsd:restriction base="xsd:NMTOKEN">
                      <xsd:enumeration value="public"/>
                      <xsd:enumeration value="protect"/>
                      <xsd:enumeration value="private"/>
                    </xsd:restriction>
                  </xsd:simpleType>
                </xsd:attribute>
              </xsd:extension>
            </xsd:complexContent>
          </xsd:complexType>
        </xsd:element>
        <xsd:element name="Integration_link" type="Unification_linkType"
maxOccurs="unbounded">
          <xsd:annotation>
            <xsd:documentation>A link to indicate the unification relationship to
share/refer a class in different models</xsd:documentation>
          </xsd:annotation>
        </xsd:element>
        <xsd:element name="Sourcecode_link" type="Unification_linkType">
```

```
                    <xsd:annotation>
                      <xsd:documentation>A link to connect the corresponding source code
to its class in design</xsd:documentation>
                    </xsd:annotation>
                  </xsd:element>
                  <xsd:element name="Abstraction_link" type="Unification_linkType"
minOccurs="0" maxOccurs="unbounded"/>
                </xsd:sequence>
              </xsd:extension>
            </xsd:complexContent>
          </xsd:complexType>
        </xsd:element>
        <xsd:element name="Association" maxOccurs="unbounded">
          <xsd:complexType>
            <xsd:complexContent>
              <xsd:extension base="AssociationType">
                <xsd:sequence>
                  <xsd:element name="Integration_link" type="Unification_linkType"
maxOccurs="unbounded"/>
                </xsd:sequence>
                <xsd:attribute name="id" type="xsd:string"/>
              </xsd:extension>
            </xsd:complexContent>
          </xsd:complexType>
        </xsd:element>
        <xsd:element name="Class_Diagram" maxOccurs="unbounded">
          <xsd:complexType>
            <xsd:sequence>
              <xsd:element name="Class" maxOccurs="unbounded">
                <xsd:annotation>
                  <xsd:documentation>Element(component) of XUM for design: UML
Class Diagram association</xsd:documentation>
                </xsd:annotation>
                <xsd:complexType>
                  <xsd:sequence>
                    <xsd:element name="Integration_link" type="Unification_linkType"
minOccurs="0">
                      <xsd:annotation>
                        <xsd:documentation>A link to indicate the unification relationship
to share/refer a class in different models</xsd:documentation>
                      </xsd:annotation>
                    </xsd:element>
                  </xsd:sequence>
                  <xsd:attribute name="name" type="xsd:string"/>
                </xsd:complexType>
              </xsd:element>
              <xsd:element name="Class_Association" maxOccurs="unbounded">
                <xsd:annotation>
                  <xsd:documentation>Element(association) of XUM for design: UML
Class Diagram association</xsd:documentation>
                </xsd:annotation>
                <xsd:complexType>
                  <xsd:complexContent>
```

```
                        <xsd:extension base="AssociationType">
                         <xsd:sequence>
                           <xsd:element name="Integration_link"
                                   type="Unification_linkType"/>
                         </xsd:sequence>
                          <xsd:attribute name="type">
                            <xsd:simpleType>
                              <xsd:restriction base="xsd:NMTOKEN">
                               <xsd:enumeration value="association"/>
                               <xsd:enumeration value="composition"/>
                               <xsd:enumeration value="generalization"/>
                               <xsd:enumeration value="dependency"/>
                              </xsd:restriction>
                            </xsd:simpleType>
                          </xsd:attribute>
                          <xsd:attribute name="client">
                            <xsd:simpleType>
                              <xsd:restriction base="xsd:NMTOKEN">
                               <xsd:enumeration value="0"/>
                               <xsd:enumeration value="1"/>
                               <xsd:enumeration value="n"/>
                               <xsd:enumeration value="0..1"/>
                               <xsd:enumeration value="0..n"/>
                               <xsd:enumeration value="1..n"/>
                              </xsd:restriction>
                            </xsd:simpleType>
                          </xsd:attribute>
                        </xsd:extension>
                      </xsd:complexContent>
                    </xsd:complexType>
                  </xsd:element>
                   <xsd:element name="Abstraction_link" type="Unification_linkType"
minOccurs="0"/>
                </xsd:sequence>
                <xsd:attribute name="dominator" type="xsd:string" use="required"/>
              </xsd:complexType>
            </xsd:element>
            <xsd:element name="Collaboration_Diagram" minOccurs="0"
maxOccurs="unbounded">
              <xsd:complexType>
                <xsd:sequence>
                  <xsd:element name="Class" maxOccurs="unbounded">
                    <xsd:annotation>
                      <xsd:documentation>Element(component) of XUM for design: UML
Collaboration Diagram association</xsd:documentation>
                    </xsd:annotation>
                    <xsd:complexType>
                      <xsd:sequence>
                        <xsd:element name="Integration_link" type="Unification_linkType">
                          <xsd:annotation>
                            <xsd:documentation>A link to indicate the unification relationship
to share/refer a class in different models</xsd:documentation>
                          </xsd:annotation>
```

```
                    </xsd:element>
                  </xsd:sequence>
                  <xsd:attribute name="name" type="xsd:string"/>
                </xsd:complexType>
              </xsd:element>
              <xsd:element name="Collaboration_Association" maxOccurs="unbounded">
                <xsd:annotation>
                  <xsd:documentation>Element(association) of XUM for design: UML
Collaboration Diagram association</xsd:documentation>
                </xsd:annotation>
                <xsd:complexType>
                  <xsd:complexContent>
                    <xsd:extension base="AssociationType">
                      <xsd:sequence>
                        <xsd:element name="Integration_link"
type="Unification_linkType"/>
                      </xsd:sequence>
                      <xsd:attribute name="sequence" type="xsd:string"/>
                      <xsd:attribute name="message" type="xsd:string"/>
                    </xsd:extension>
                  </xsd:complexContent>
                </xsd:complexType>
              </xsd:element>
              <xsd:element name="Abstraction_link" type="Unification_linkType"
minOccurs="0"/>
            </xsd:sequence>
            <xsd:attribute name="dominator" type="xsd:string" use="required"/>
          </xsd:complexType>
        </xsd:element>
        <xsd:element name="Design_Pattern" minOccurs="0" maxOccurs="unbounded">
          <xsd:annotation>
            <xsd:documentation>Element(component) of XUM for design: Design
Pattern</xsd:documentation>
          </xsd:annotation>
          <xsd:complexType>
            <xsd:complexContent>
              <xsd:extension base="ComponentType">
                <xsd:sequence>
                  <xsd:element name="Participation" maxOccurs="unbounded">
                    <xsd:annotation>
                      <xsd:documentation>Element(component) of XUM for design:
Design Pattern: Participation</xsd:documentation>
                    </xsd:annotation>
                    <xsd:complexType>
                      <xsd:complexContent>
                        <xsd:extension base="ComponentType">
                          <xsd:sequence>
                            <xsd:element name="Integration_link"
type="Unification_linkType">
                              <xsd:annotation>
                                <xsd:documentation>A link to indicate the unification
relationship to share/refer a common class in different models</xsd:documentation>
                              </xsd:annotation>
```

```
                </xsd:element>
              </xsd:sequence>
              <xsd:attribute name="role" type="xsd:string"/>
            </xsd:extension>
          </xsd:complexContent>
        </xsd:complexType>
      </xsd:element>
      <xsd:element name="Structure" maxOccurs="unbounded">
        <xsd:annotation>
          <xsd:documentation>Element(association) of XUM for design: Design
Pattern structure association</xsd:documentation>
        </xsd:annotation>
        <xsd:complexType>
          <xsd:complexContent>
            <xsd:extension base="AssociationType">
              <xsd:sequence>
                <xsd:element name="Integration_link"
type="Unification_linkType"/>
              </xsd:sequence>
              <xsd:attribute name="type">
                <xsd:simpleType>
                  <xsd:restriction base="xsd:NMTOKEN">
                    <xsd:enumeration value="association"/>
                    <xsd:enumeration value="composition"/>
                    <xsd:enumeration value="generalization"/>
                    <xsd:enumeration value="dependency"/>
                  </xsd:restriction>
                </xsd:simpleType>
              </xsd:attribute>
              <xsd:attribute name="client">
                <xsd:simpleType>
                  <xsd:restriction base="xsd:NMTOKEN">
                    <xsd:enumeration value="0"/>
                    <xsd:enumeration value="1"/>
                    <xsd:enumeration value="n"/>
                    <xsd:enumeration value="0..1"/>
                    <xsd:enumeration value="0..n"/>
                    <xsd:enumeration value="1..n"/>
                  </xsd:restriction>
                </xsd:simpleType>
              </xsd:attribute>
            </xsd:extension>
          </xsd:complexContent>
        </xsd:complexType>
      </xsd:element>
      <xsd:element name="Collaboration" maxOccurs="unbounded">
        <xsd:annotation>
          <xsd:documentation>Element(association) of XUM for design: Design
Pattern collaboration association</xsd:documentation>
        </xsd:annotation>
        <xsd:complexType>
          <xsd:complexContent>
            <xsd:extension base="AssociationType">
```

```
                    <xsd:sequence>
                      <xsd:element name="Integration_link"
type="Unification_linkType"/>
                    </xsd:sequence>
                    <xsd:attribute name="sequence" type="xsd:string"/>
                    <xsd:attribute name="message" type="xsd:string"/>
                  </xsd:extension>
                </xsd:complexContent>
              </xsd:complexType>
            </xsd:element>
            <xsd:element name="Abstraction_link" type="Unification_linkType"
minOccurs="0"/>
          </xsd:sequence>
          <xsd:attribute name="dominator" type="xsd:string"/>
        </xsd:extension>
      </xsd:complexContent>
    </xsd:complexType>
   </xsd:element>
  </xsd:sequence>
 </xsd:complexType>
</xsd:element>
<xsd:element name="Implementation">
  <xsd:complexType>
    <xsd:sequence>
      <xsd:element name="Framework" minOccurs="0" maxOccurs="unbounded">
        <xsd:complexType>
          <xsd:sequence>
            <xsd:element name="Sourcecode" maxOccurs="unbounded">
              <xsd:annotation>
                <xsd:documentation>Element(component) of XUM for implementation:
source code </xsd:documentation>
              </xsd:annotation>
              <xsd:complexType>
                <xsd:sequence>
                  <xsd:element name="Sourcecode_link" type="Unification_linkType">
                    <xsd:annotation>
                      <xsd:documentation>A link to connect a class in the design to its
corresponding codes</xsd:documentation>
                    </xsd:annotation>
                  </xsd:element>
                </xsd:sequence>
              </xsd:complexType>
            </xsd:element>
          </xsd:sequence>
        </xsd:complexType>
      </xsd:element>
    </xsd:sequence>
  </xsd:complexType>
</xsd:element>
   </xsd:sequence>
  </xsd:complexType>
 </xsd:element>
</xsd:schema>
```

Chapter VI

Migrating Legacy System to the Web: A Business Process Reengineering Oriented Approach

Lerina Aversano, Gerardo Canfora, Andrea De Lucia
Research Centre on Software Technology (RCOST)
University of Sannio, Italy

The Internet is an extremely important new technology that is changing the way in which organizations conduct their business and interact with their partners and customers. To take advantage of the Internet open architecture, most companies are applying business reengineering with the aim of moving from hierarchical centralized structures to networked decentralized business units cooperating with one another. As a consequence, the way in which software information systems are conceived, designed, and built is changing too. Monolithic, mainframe-based systems are being replaced by distributed, Web-centric, component-based systems with an open architecture.

Ideally, business process reengineering should entail the adoption of new software systems designed to satisfy the new needs of the redesigned business. However, economic and technical constraints make it impossible in most cases to discard the existing and legacy systems and develop replacement systems from scratch. Therefore, legacy system migration strategies are often preferred to replacement. This entails that a balance must be struck between the constraints imposed by the existing legacy systems and the opportunities offered by the reengineering of the business processes.

This chapter discusses a strategy for migrating business processes and the supporting legacy systems to an open, networked, Web-centric architecture. The overall strategy comprises modelling the existing business processes and assessing the business and technical value of the supporting software systems. A decisional framework helps software management to make informed decisions. This is followed by the migration of the legacy systems, which can be in turn enacted with different approaches. The chapter discusses a short-term migration approach and applies it to an industrial pilot project.

INTRODUCTION

Convergence between telecommunications and computing and the explosion of the Internet suggest new ways of conceiving, designing, and running businesses and enterprises. More and more companies are moving towards a virtual organization model, where independent institutions, departments, and groups of specialized individuals converge in a temporary network with the aim of utilizing a competitive advantage or solving a specific problem.

Information and communication technology is a primary enabler of virtual organizations, as people and institutions in a network make substantially more use of computer-mediated channels than physical presence to interact and cooperate in order to achieve their objectives. However, technology is not the only factor: taking advantage of the Internet and its open architecture requires that the way in which business processes are organized and enacted be profoundly changed. Business Process Reengineering (BPR) is defined as "the fundamental rethinking and radical redesign of business processes to achieve significant improvements of the performances, such as cost, quality, service, and speed" (Hammer & Champy, 1993). Most BPR projects aim at converting business organizations from hierarchical centralized structures to networked decentralized business units cooperating with one another. This conversion is assuming a strategic relevance as the Internet is radically changing business processes, not only because they are purposely reengineered, but also because the Internet and, in general, the information and communication technology, offer clients and customers more convenient means of fulfilling their requirements.

Current business processes have been profoundly fitted to the available hardware and software. The technologies involved in process execution impact the way businesses are conceived and conducted. Abstractly, reengineering business processes should entail discarding the existing and legacy systems to develop new software systems that meet the new business needs. This is superficially attractive and humanly appealing. However, in most cases, legacy systems cannot be simply discarded because they are crucial systems to the business they support (most legacy systems hold terabytes of live data) and encapsulate a great deal of knowledge and expertise of the application domain. Sometimes, the legacy code is the only place where domain knowledge and business rules are recorded, and this entails that even the development of a new replacement system may have to rely on knowledge that is encapsulated in the old system. In summary, existing systems are the result of large investments and represent a patrimony to be salvaged (Bennett, 1995). In addition, developing a replacement system from scratch often requires excessive resources and entails unaffordable costs and risks. Even where these resources are available, it would take years before the functions provided by a legacy system could be taken up by new reliable components. Legacy system migration strategies are often preferred to replacement (Brodie & Stonebaker, 1995). Therefore, to satisfy the goals of a BPR project it is necessary to work intensively to search for a trade-off between the constraints of existing legacy systems and the opportune BPR strategy.

In this chapter, we discuss a strategy for migrating business processes and the supporting legacy systems to an open, networked, Web-centric architecture. The overall strategy comprises modelling the existing business processes and assessing the business and quality value of the supporting software systems. This is followed by the migration of the legacy systems, which can in turn be enacted with different strategies. The initial step consists of understanding and modelling the business processes together with the involved documents and software systems. The analysis of the existing processes is required to get an inventory of the activity performed, compare them with best practices, and redesign and/or reengineer them.

Jacobson, Ericsson, and Jacobson (1995) introduce the term *process reverse engineering* to denote the activities aimed at understanding and modelling existing business processes and define a two-step method:

- use case modeling, that produces a model of the existing business processes in terms of actors and use cases; and
- object modeling, that produces an object model of the processes.

The aim is to identify a basic process model in order to prioritize the processes that are most critical for further investigations. Therefore, the model produced tends to be at a very high level of abstraction; the authors recommend that "you should be careful not to create models that are too detailed" (Jacobson et al., 1995). In addition, the authors do not give insights on how the information and data needed to create the models should be collected.

We have developed a method for business process understanding and modelling and have applied it in two different areas, redesigning administrative processes in the public organizations (Aversano, Canfora, De Lucia, & Gallucci, 2002) and providing automatic support for the management of maintenance processes within a virtual software factory (Aversano, Canfora, & Stefanucci,

2001b; Aversano, De Lucia, & Stefanucci, 2001c). The main difference with the work by Jacobson et al. (1995) is that we aim at producing process models that are sufficiently detailed to be readily used for process automation with a workflow management system (Workflow Management Coalition, 1994). The method addresses the modelling of a business process from different perspectives:

- the activities comprised in the process, the flow of control among them, and the decisions involved;
- the roles that take part in the process; and
- the flow and the structure of information and documents processed and/or produced in the process.

The method also addresses the analysis and the assessment of the software systems that support and automate the process.

Different formalisms and languages for process modelling have been proposed in the literature (Casati, Ceri, Pernici, & Pozzi, 1995; Van der Aalst, 1998; Winograd & Flores, 1986). Our approach is different: we do not introduce a new modelling language, rather we use the standard UML (Booch, Rumbaugh, & Jacobs, 1999), which has also been used as a process and workflow modelling language also (Cantone, 2000; Loops & Allweyer, 1998). In particular, Loops and Allweyer (1998) show how a suitable workflow model can be achieved through the joint use of different UML diagrams (activity diagrams, use-case diagrams, and interaction diagrams), although some deficiencies are pointed out. In our opinion, a reasonable high-level workflow model can be produced using a combination of UML diagrams, and refined using the specific process definition language constructs of the selected workflow management system.

There are also several alternative formalisms to approach the modelling of the set of documents involved in a process and the actors producing/processing them. An example is the Information Control Nets (Salminen, Lytikainen, & Tiitinen, 2000), which provides all the characteristics and notations necessary to represent, describe, and automate the flow of documents. We use UML diagrams to model the structure, the information content, and the flows of documents, too. An advantage is that the processes and the associated documents are represented in an integrated way.

Our overall approach for business process understanding and modelling consists of five main phases, each comprising several steps: Phases 1 and 2 define the context and collect the data needed to describe the existing processes, with related activities, documents, and software systems; Phases 3, 4, and 5 are executed in parallel and focus on workflow modelling, document modelling, and legacy system analysis and assessment, respectively. Figure 1 summarizes the main aspects of the different phases. The first four phases of our approach are discussed in detail in references (Aversano et al., 2002; Aversano et al., 2001b). In this chapter we focus on the final phase and discuss how the business and technical value of a legacy system can be assessed and how the results of the assessment can be used to select a system migration strategy.

Figure 1: Guidelines description.

Context definition. The first phase consists of two steps: scope definition and process map definition. The goal of the scope definition step is the identification of two key elements: the context of the analyzed organization (i.e., the units it comprises and the other entities it interacts with) and the products (i.e., the objects handled within the processes of the organization). In the process map definition step, a description of the process at different levels of abstraction is produced according to a top-down approach. Defining a process map helps to identify the key processes, the involved documents, and software systems.
Data collection and process description. The goal of this phase is to collect all information needed to describe and assess the existing processes and the involved documents and systems. This information can be achieved in different ways, for example, through observations, questionnaires, and interviews conducted with process owners and key users. The collected information about workflows, documents, and software systems are included in a structured process description document.
Workflow modelling. Workflow modelling involves two separate steps: activity map definition and activity documentation. The first step aims to produce a semi-formal graphical model of the analyzed process. We use UML activity diagrams to model the flow of the process activities, including decisions and synchronizations, use cases to model organizational aspects, i.e., which actors (roles) participate in which use case (activity or group of activities), and interaction (sequence and collaboration) diagrams to depict dynamic aspects within a use case. In the activity documentation step, the workflow model produced in the previous step is completed with detailed documentation.
Document modelling. This phase produces a model of the content, structure, and mutual relationships of the documents involved in the workflow. It is based on two steps: document class definition and document life-cycle modeling. Initially, the documents are partitioned into classes on the basis of their content, and the relationships existing between document classes are identified. The result is a document-relationships model modelled through a UML class diagram, where nodes represent document classes and edges depict mutual relationships. The second step consists of describing the life cycle for each document class, i.e., its dynamic behavior, through UML state diagram.
Legacy systems analysis and assessment. This phase aims to assess the existing software systems involved in the business process to identify the most feasible Web migration strategy. Usually, a legacy system is assessed from two points of views: a business dimension and a technical dimension. Most of the information for the evaluation of the business value is collected through interviews and questionnaires. The technical value of a legacy system can be assessed through different quality attributes, such as the obsolescence of the hardware/software platforms, the level of decomposability, the maintainability, and the deterioration. This information can be achieved by combining interviews and questionnaires with the analysis of the legacy source code and of the available documentation.

The chapter is organized as follows. The next section discusses the assessment of legacy systems and presents decisional frameworks to guide and support management in making decisions about a system. The decisional frameworks presented analyze the legacy systems from both the business and the technical points of view. Then, the chapter introduces a strategy to migrate legacy systems as a consequence of a BPR project and discusses the main enabling technologies. The application of the strategy in an industrial pilot project is then discussed. Finally, the chapter summarizes the work and gives concluding remarks.

MAKING DECISIONS ABOUT LEGACY SYSTEMS

There are a number of options available in managing legacy systems. Typical solutions include: discarding the legacy system and building a replacement system; freezing the system and using it as a component of a new larger system; or modifying the system to give it another lease on life. Modifications may range from a simplification of the system (reduction of size and complexity) to preventive maintenance (redocumentation, restructuring, and reengineering) or even to extraordinary processes of adaptive maintenance (interface modification, wrapping, and migration) (Canfora & Cimitile, 1997; De Lucia, Fasolino, & Pompella, 2001; Pigoski, 1997). These alternatives are not mutually exclusive, and the decision on which approach, or combination of approaches, is the most suitable for any particular legacy systems must be made based on an assessment of the technical and business value of the system.

In particular, reengineering business processes to take advantage of the open architecture of the Internet is likely to entail a major process of adaptations of the legacy systems to a Web-centric architecture, or even the replacement/redevelopment of the systems. It is widely documented in the literature (Sneed, 1995; Sneed, 1999) that migrating a legacy system towards a new networked platform entails costs and risks that depend on the characteristics of both the architecture of the source system and the target platform. As a consequence, to perform a BPR project, it is necessary to take into account the possible impact of the process redesign on the existing legacy systems. It is important to evaluate the costs of the different options available and, in particular, replacement with standardized packages (when available) redevelopment (possibly after a reverse engineering stage), or migration.

Several authors have identified possible alternatives for dealing with legacy systems and have proposed decision frameworks to select among the alternatives. Bennett, Ramage and Munro (1999) identify six strategies to deal with a legacy system: discard, wrap, outsource, freeze, carry on with maintenance and support, and reverse engineering. They stress the need to model the business strategy of an organization from a top-down perspective, including many stakeholders to make informed decisions about legacy systems. They introduce a two-phase model, called Software as a Business Asset (SABA), that uses an organizational scenario tool to generate scenarios for the organization's future and a technology scenario tool to produce a prioritized set of solutions for the legacy system. Prioritization of solutions is achieved by comparing the current (legacy) system with the systems required by each scenario generated by the organizational scenario tool.

Sneed (1995) suggests that five steps should be considered when planning a reengineering project: project justification, which entails determining the degree to which the business value of the system will be enhanced; portfolio analysis, which consists of prioritizing the applications to be reengineered based on their technical quality and business value; cost estimation, which is the estimation of the costs of the project; cost-benefit analysis, in which costs and expected returns are compared; and contracting, which entails the identification of tasks and the distribution of effort.

In general, decision frameworks require that a legacy system be assessed from two points of views: a business dimension and a technical dimension (Bennett et al., 1999; Sneed, 1995; Verdugo, 1988).

The data required for the evaluation of the business value is collected mainly through observations, interviews and questionnaires (see the second phase of the method in Figure 1). This information measures the complexity of the business processes and administrative rules that a system (or system's component) implements and their relevance to achieving business competitiveness.

The technical value of a legacy system can be assessed through different quality attributes, such as the obsolescence of the hardware/software platforms, the level of decomposability, the maintainability, and the deterioration (Canfora & Cimitile, 1997; De Lucia et al., 2001). The obsolescence measures the aging of the existing systems due to the progresses of software and data engineering and the evolution of hardware/software platform. Obsolescence induces a cost that results from not taking advantage of modern methods and technologies that reduce the burden of managing and maintaining existing systems. The deterioration measures the decrease of the maintainability of existing systems due to the maintenance operations the system has undergone in its life cycles. Deterioration directly affects the cost of managing and maintaining a system. The decomposability measures the identifiability and independence of the main components of a system. This information can be obtained by combining interviews and questionnaires with the analysis of the legacy source code and of the available documentation.

A typical decision framework for the management of legacy systems based on four main alternatives is shown in Figure 2 (De Lucia et al., 2001). The ordinary maintenance decision is generally adopted in the case of a system with a good technical and business value; this alternative entails only ordinary interventions on the system, aiming at adding new functionality or resolving specific problems. On the other hand, the elimination of the system and replacement with a new one, developed adhoc or acquired from the market, is generally forced by the low technical value and low business value of the existing system. The evolution alternative aims to evolve the system by providing it with new functionality or adapting the existing ones in order to improve the business value of the system, while maintaining the high technical value. The last option is the reengineering / migration alternative; in this case, the goal is more complex and entails moving the legacy system toward a new and more flexible environment, while retaining the original system's data and functionality; typically, reverse engineering, restructuring, and architecture transformation interventions are suggested to improve software technical attributes in a system with high business value.

In De Lucia et al. (2001) such a framework is refined with guidelines and rules used to select the particular type of maintenance intervention on the basis of the specific values of the business and technical attributes of the assessment model. In this chapter, we assess the technical quality of a legacy system by considering the obsolescence and the decomposability level. In particular, we focus on making decisions on the actions to perform on a legacy system as a consequence of a BPR project aimed at taking advantage of the Internet. We assume that the obsolescence of the system is high, and therefore extraordinary maintenance is required to adapt it to the Web. Accordingly, the decision about the particular type of intervention to take will be made based on the decomposability and business value of the system.

Figure 3 shows a customized decisional framework that takes into account the considerations above; some of the options in Figure 2 are not available any more, as the system must necessarily change; i.e., the ordinary maintenance option does not apply, and evolution assumes the form of a componentization of the system, in

Figure 2: Typical decisional framework.

Technical value	Low (Business value)	High (Business value)
High	Evolution	Ordinary Maintenance
Low	Elimination / Replacement	Redevelopment / Reengineering / Migration

view of an incremental migration. In addition, the main technical factor affecting the possibility of renewing the legacy system into the Web-based infrastructure of the reengineered business process is the system decomposability, which therefore is assumed as a decision dimension rather than as a component of the technical value.

Two different kinds of decomposability can be considered:

- *vertical decomposability*, which refers to the possibility of decomposing a system into major architectural layers; and

Figure 3: A customized decisional framework.

Decomposability value	Low (*Business value*)	High (*Business value*)
High	Component based Incremental Migration	Short term Migration
Low	Elimination / Replacement	Reengineering / Redevelopment

Figure 4: Vertical and horizontal decomposability.

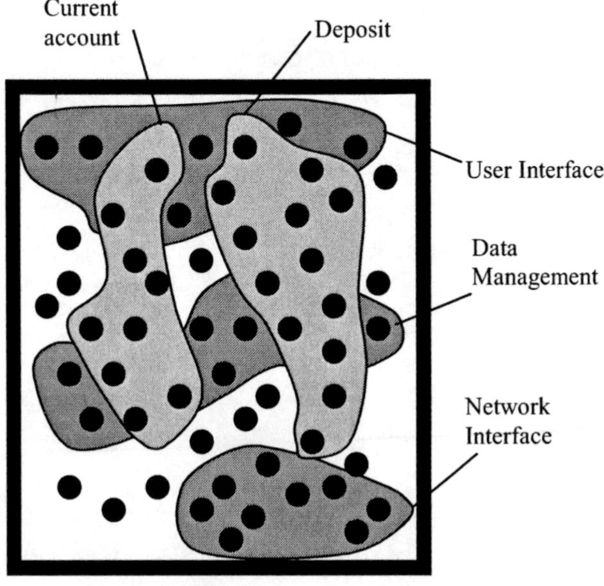

- *horizontal decomposability*, which accounts for the possibility of decomposing a legacy system into independent and cooperating business components.

Figure 4 illustrates the vertical and horizontal decomposability definitions. The system is depicted as a set of code components, such as procedures, functions, files, objects, etc. Vertical decomposability refers to the identification of layers of components (for example, the user interface, the data management, and the network interface layers), whereas the horizontal decomposability addresses the identification of groups of components, possibly across layers, that implement domain concepts (for example, a current account or deposit in a banking system).

In particular, concerning the vertical decomposability, Brodie and Stonebaker (1995) refer that a software system can be considered as having three types of components: *interface components*, *application logic components*, and *database components* (we will refer to application logic and database components as server components). Depending on how separated and well identified are these components, the architecture of a legacy system can be *decomposable*, *semidecomposable*, or *nondecomposable*. In a decomposable system the application logic components are independent of each other and interact with the database components and potentially with their own user and system interfaces. In a semidecomposable system, only interfaces are separate modules, while application logic components and database services are not separated. A nondecomposable system has the worst architecture; i.e., the system is a black box with no separated components.

The decision about the intervention to take on the legacy system with a high business value is mainly driven by the vertical decomposability of the system. If the vertical decomposability value is sufficiently high, i.e., the system is decomposable or semidecomposable in the Brodie and Stonebaker terminology, the best strategy is a short-term migration of the system, achieved through wrapping the application logic and/or data management functions (that define the server tier) and reengineering/redeveloping the user interface to a Web-centric style. This strategy represents a good alternative in the case of a nondecomposable system, provided that the costs and risks of its decomposition are affordable (Canfora, De Lucia, & DiLucca, 1999; Sneed, 1995); techniques and tools for decomposing monolithic systems are discussed in Canfora, Cimitile, De Lucia, and DiLucca (2000). If the decomposability level of a legacy system with high business value is too low, the complete reengineering / redevelopment alternative has to be preferred, as the legacy system can still be used as the basis to abstract the specification and design documents by reverse engineering.

For legacy systems with a low business value, the decision about the intervention to take is mainly driven by the horizontal decomposability. In particular, if the horizontal decomposability value is high, it is possible to again use the framework in Figure 3 to make different decisions for the different business components identified in the system, depending on the business value and the vertical decomposability of each component. This provides the basis for a component-based incremental migration of the legacy system. Whenever both the business value and the decomposability of a legacy system are low, the only possible option is elimination / replacement. Indeed, in that case, there are no adequate business preconditions to evolve the existing system into a Web-based one.

As far as the horizontal decomposability is concerned, several methods have been presented in the literature to identify business components (either in the form of objects or functional abstractions) and to evaluate the effort required to decouple them, i.e., an estimation of the horizontal decomposability of the system. Sneed and Nyary (1994) propose three different approaches—*procedural*, *functional*, and *data type*—for remodularizing and *downsizing* legacy systems from a centralized mainframe environment to a distributed client-server platform. The procedural approach views a program as a directed graph with the decisions as nodes and the branches as edges; complex graphs are split into subgraphs by finding the points of minimum interconnections. Other procedural approaches exploit program slicing techniques for program restructuring (Kim & Kwon, 1994) or to identify reusable functions (Lanubile & Visaggio, 1997). The functional approach assumes that a program has been designed according to a functional decomposition; the program can be viewed as a hierarchy of superordinate control nodes and subordinate elementary nodes, and each business rule can be mapped onto one or more control nodes. Examples of remodularization techniques based on the functional approach can be found in Cimitile and Visaggio (1995), Markosian, Newcomb, Brand, Burson, and Kitzmiller (1994), and Müller, Orgun, Tilley, and Uhl (1993). The data type approach views a program as a set of cooperative processing objects; modules are constructed by clustering together the set of operations upon a given data type or entity. Several methods have been proposed in the literature for identifying objects in legacy systems; most methods share the idea of representing a system in the form of a bipartite graph, where nodes model procedural components and data stores at different levels of granularity, while edges depict relations between them. These methods search for notable subgraphs in the bipartite graph (e.g., strongly connected subgraphs) (Dunn & Knight, 1993; Liu & Wilde, 1990; Livadas & Johnson, 1994), or exploit different algorithms to cluster the business components (Wiggerts, 1997), based on metrics (Canfora et al., 1996; Cimitile, De Lucia, DiLucca, & Fasolina, 1999), data mining (Montes de Oca & Carver, 1998), and mathematical concept analysis (Lindig & Snelting, 1997; Siff & Reps, 1999; Canfora, Cimitile, De Lucia, & DiLucca, 2001).

A MIGRATION STRATEGY

The migration of a legacy system entails the reuse of the system components while moving the system toward newer and more modern technology infrastructure. Brodie and Stonebraker (1995) propose an incremental approach named Chicken Little, based on 11 steps to migrate a legacy information system using gateways. Each step requires a relatively small resource allocation and takes a short time; it produces a specific, small result toward the desired goal corresponding to an increment. The Chicken Little steps are: analyzing the legacy system, decomposing the legacy system structure, designing the target interfaces, designing the target application, designing the target database, installing the target environment, creating and installing the necessary gateway, migrating the legacy database, migrating the legacy applications, migrating the legacy interfaces, and cutting over to the target system. Using this strategy, it is possible to control the risk involved and determine the size of each part of the legacy system to be migrated. If a step fails, it is not necessary to repair the entire process but only to restore the failed step. A different approach proposed by Wu et al. (1997) is the Butterfly methodology that eliminates the needs to access both the legacy and new database during the migration process, thus avoiding the complexity induced by the introduction of gateways to maintain the consistency of the data.

The Chicken Little and the Butterfly methodologies aim at migrating a legacy system mainly based on its vertical decomposability. Migration strategies have also been proposed that takes into account the horizontal decomposability of a legacy system. In Canfora et al. (1999) a strategy for incrementally migrating legacy systems to object-oriented platforms is presented. The process consists of six sequential phases: static analysis of legacy code, decomposing interactive programs, abstracting an object-oriented model, packing the identified object methods into new programs, encapsulating existing objects using wrappers, and incrementally replacing wrapped objects. Wrapping is the core of the migration strategy: it makes new systems able to exploit existing resources, thus allowing an incremental and selective replacement of the identified objects. Serrano, Montes de Oca, and Carter (1999) propose a similar approach. The main difference between the two migration strategies is the method used to identify objects. Serrano et al. (1999) exploit data mining techniques (Montes de Oca & Carver, 1998), while Canfora et al. (1999) use a clustering method based on design metrics (Cimitile et al., 1999).

In this chapter we propose a Web-centric, short-term migration strategy based on vertical decomposability and on the use of wrappers. As discussed in the previous section, this strategy applies to legacy systems with high business value and a high vertical decomposability level. Figure 5 shows the main phases of the migration process. The first phase is the decomposition and restructuring of the legacy system to a client-server style, where the user interface (client side) controls

Figure 5: Short-term migration strategy.

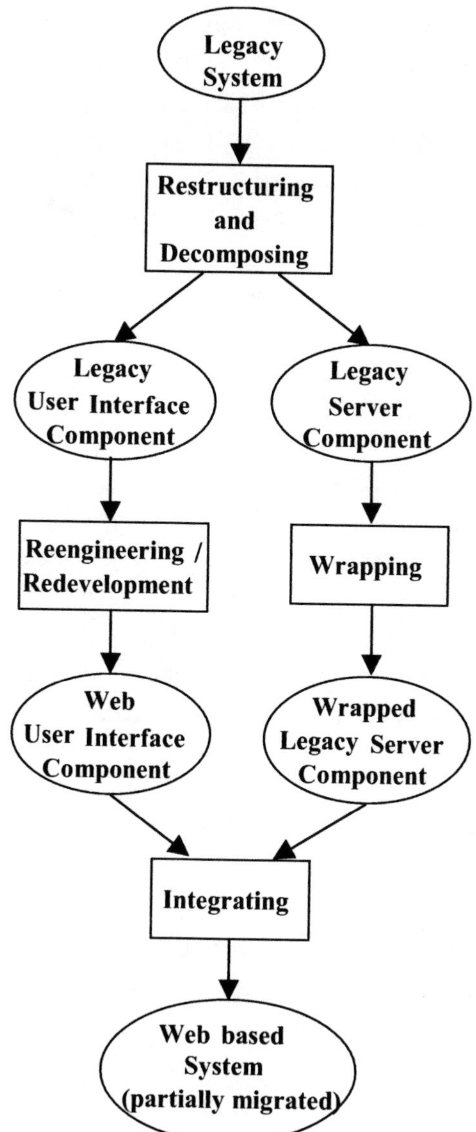

the execution of the application logic and database components (server side). The complexity of this phase depends on the decomposability of the system. Nondecomposable systems are the most difficult to be restructured, because decoupling the user interface from the application logic and database components can be very challenging and risky; slicing algorithms are required to identify the statements that implements the user-interface component, while restructuring the legacy programs to a client server style is human intensive (Canfora et al., 2000).

However, even migrating semidecomposable or decomposable systems might not be trivial.

Once the legacy system has been restructured to a client-server style, if needed, two phases can be conducted in parallel, namely, the user-interface reengineering and the wrapping of the server tier. The first phase aims at reengineering/redeveloping the user interface of the system by exploiting Web technologies. The design of the new interface is driven by the analysis of the legacy user interface and by the new modalities of accessing the system derived from the BPR goals (Sneed, 2000). Wrapping the server tier is the key to reusing a legacy system with a high business value. However, the legacy programs might still require some adaptations to encapsulate them into object wrappers. A wrapper can be implemented using different technologies, including distributed component middleware; the main goal is to transform a message sent by the user-interface component into an invocation of the corresponding legacy program with the appropriate parameters. The final step is the integration and testing of the new Web-based application.

It is worth noting that once the Web-based migration of the legacy system is completed, the management of the new system will be based on the general decision framework in Figure 2. In particular, the main factor to consider will be the business value, at least until the technical value of the system is adequate to meet the evolving business needs. In other words, the migrated system will undergo periodical phases of evolution—to keep its functionality in line with the evolving business goals—interleaved with the ordinary maintenance of the system. Of course, a major shift in the technological business needs will entail a rapid increase of the obsolescence of the system, thus moving the management of the system to the decision framework of Figure 3.

Decomposition and Restructuring

The restructuring and decomposition phase is preliminary to the Web-based reengineering of the user interface and wrapping of the server side of the legacy system. The method to be adopted depends on the compliance of the system to a client-server style and on its vertical decomposability level. A semidecomposable or decomposable system is compliant to a client-server style if the user-interface component has the control and communicates with the other components through invocations; this is typical of legacy systems that have a graphical user-interface (GUI). For these systems, the restructuring actions to be taken depend on the coupling between the user-interface components and the other components. If all the data are exchanged through the calls between the client side and the application logic components, there is no need for restructuring. However, if the client contains instructions to access the database component, it might be required to extract these code fragments from the user-interface component and encapsulate them into separate components of the application logic layer; alternatively, these instructions can be left in the user-interface and reimplemented according to the method and the technologies adopted in the user interface reengineering phase. If the client and server components exchange some data through global data areas, data flow analysis and restructuring techniques are required to transform the global data into parameters to be exchanged across calls (Cimitile, DeCarlin, & De Lucia, 1998; Markosian et al., 1994; Sneed & Nyary, 1994).

However, a legacy system can be decomposable or semidecomposable, although it is not compliant to a client-server style. For example, legacy systems developed in languages such as COBOL or RPG often have a well-defined and identifiable character-based user interface; however, generally, the application logic component has the control and invokes the user interface components. In this case, there is a need for shifting the dialogue initiative between the user interface and application logic components (Moore & Moshkina, 2000). The same problem has to be tackled for nondecomposable systems with affordable restructuring costs and risks. In Canfora et al. (2000), a method has been presented to identify the code implementing the user interface of these types of legacy systems and reorganize the legacy programs to a client-server style. The reorganization may entail decomposing and restructuring a program whenever it comprises both components that realize the user interface and components that implement the application logic and database services.

One of the main difficulties that prevent the migration of these systems is the interleaving in the same subroutine of statements implementing both types of components; a particular case is when subroutines implementing database components (and that should then be migrated to a server) contain calls to subroutines implementing user interface components (that should be migrated to the client). Figure 6(a) shows a typical call graph of a legacy program: grey nodes (*s2*, *s8*, and *s13*) represent subroutines containing I/O instructions or calls to user-interface components, while light grey nodes (*s*, *s1*, *s3*, and *s4*) represent subroutines containing accesses to the database, business rules (application logic), and calls to subroutines implementing user-interface components. White components only contain database accesses and/or business rules. As some calls are directed from the subroutines implementing database and application logic components to the user-interface components, the program is not in a client-server style. To make the program a client-server style, the code fragments implementing database and application logic components have to be extracted from the light grey subroutines and encapsulated into separated subroutines. Figure 6(b) shows the same program restructured to a client-server style, where the result of the extraction of the code fragments implementing database and application logic components from the subroutines *s*, *s1*, *s3*, and *s4* has produced the new subroutines *s14*, *s15*, *s16*, *s17*, and *s18*. It is worth noting that the subroutines *s*, *s1*, *s3*, and *s4* are now colored grey, as they only contribute to implement the user-interface component due to the calls to the subroutines *s2*, *s8*, and *s13*. The new call graph can be decomposed by an ideal borderline in two parts: the subroutines of the client component (in grey) are in the higher part of the call hierarchy, while the lower part contains the subroutines of the server components (in white). The border line crosses all the edges corresponding to calls between subroutines in the client component and subroutines in the server components; these calls, depicted with the dashed lines, will have to be converted into service requests.

Canfora et al. (2000) have proposed an interprocedural slicing algorithm based on control dependencies to identify the statements implementing the user-interface components. These statements are identified by traversing the control dependencies of the legacy programs backward starting from the I/O statements. The authors also propose an interactive tool to restructure the logic of each program to a client-server style. The tool automatically identifies database accesses that need to be extracted from the user-interface part and encapsulated into separate subroutine. The tool also allows a software engineer to select code fragments implementing application logic components and automatically encapsulates them into separate subroutines; the user is only asked to assign a name to the subroutine. Extracting these subroutines allows restructuring of the program with a call graph such as the one in Figure 6(a) to a client-server style with a call graph similar to the one in Figure 6(b). Once a program has been decomposed and restructured in this way, it is ready for migration to a Web architecture. The user interface will be reengineered or redeveloped using Web technology, while the subroutines of the server component invoked by the legacy user interface will be wrapped. These two issues will be discussed further in the next sections.

User Interface Reengineering

User interface reengineering entails using reverse engineering to abstract a user interface conceptual model and forward engineering to reimplement the user

Figure 6: Call graph decomposition.

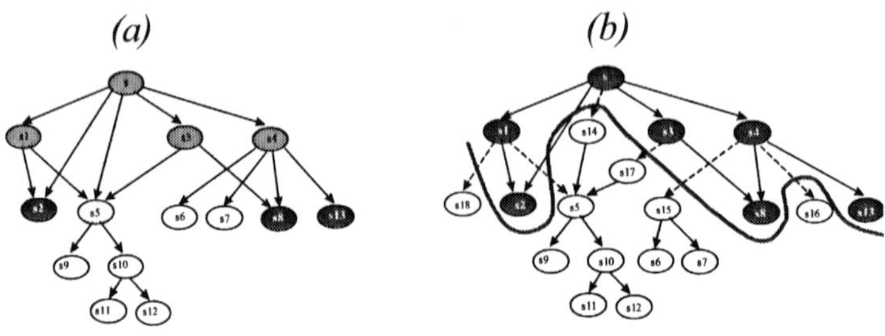

interface using Web-based technologies. The maturity of the Web offers different solutions to reimplement the user interface of the system. Several scripting languages, such as VBScript and JavaScript, have been introduced to enhance HTML and quickly develop interactive Web pages. Scripts can be executed on the client to dynamically modify the user interface, driven by the occurrence of particular events. Scripts can also be executed on the Web server, in particular to generate dynamic HTML pages from the data received from the interaction with external resources, such as database and applications.

Concerning the reengineering methodologies, several approaches to the problem of user interface reengineering have been presented in the literature, that can be customized for the Web. Merlo et al. (1995) propose a technique for reengineering CICS-based user interfaces in COBOL programs into graphical interfaces for client-server architectures. Character-based interface components are extracted and converted into specifications in the Abstract User Interface Description Language (AUIDL); then they are reengineered into graphical AUIDL specifications and used to generate the graphical user interfaces. The authors outline the need for investigating slicing and dependence analysis to integrate the new interface code into the original system. Also, this approach could be extended and used to convert a character-based user interface into a Web-based user interface. Van Sickle, Liu, and Ballantyne (1993) propose a technique for converting user interface components in large minicomputer applications, written in COBOL, to run under CICS on an IBM mainframe.

A more general approach, called MORPH, has been proposed in references (Moore, 1998; Moore & Moshkina, 2000; Moore & Rugaber, 1997). MORPH exploits a knowledge base for representing an abstract model of the interaction objects detected in the code of the legacy system. Transformation rules are exploited to restructure and transform the original abstract model into a new abstract model used to generate the target graphical user interface. MORPH was originally developed for reengineering character-based user interfaces to graphical user interfaces; it has also been used to reengineer graphical user interfaces from one platform to another (Moore & Rugaber, 1997) and character-based user interfaces to Web interfaces (Moore & Moshkina, 2000). We have also used these guidelines to reengineer graphical user interfaces to Web interfaces (Aversano, Cimitile, Canfora, & DeLucia, 2001a).

The MORPH method entails three steps, named detection, representation, and transformation. In the first step, a static analysis is conducted to identify and extract the user interface implementation patterns from the source code. The representation step aims at building a hierarchical abstract model where the identified user interface coding patterns are the leaves and higher level conceptual interaction tasks and attributes are abstracted from the lower level patterns. This abstract model is stored in the MORPH knowledge base. The final step defines a set of transformation rules used to move the abstract model into a concrete implementation with a particular GUI technology.

The transformation of the legacy user interface into the new Web interface should also be driven by the business needs. For example, if the new interface has to be used by the old users, a goal could be minimizing the need for retraining. In this case, in addition to establishing a one-to-one correspondence between the legacy and the new interaction objects, the mapping should be built maintaining a correspondence between the legacy panels and the new Web pages and forms (Aversano et al., 2001a). However, in many cases the BPR process introduces new roles (for example, a legacy system in the reengineered process might be required to be accessed directly by a customer); in this case, the new interface must be radically redesigned to meet the new user's needs (Sneed, 2000).

Wrapping and Integration

The concept of wrapping is not new, as Dietrich introduced it in 1989 (Dietrich, 1989). However, in the last years the diffusion of the wrapping technology has had a significant impulse as a consequence of the use of Internet as a means of communication between remote programs (Zou & Kontogiannis, 1999). In particular, CORBA, DCOM and COM+ are the major middleware architectures for web based applications requiring access to remote objects.

According to Mowbray and Zahari (1994), wrapping is the practice for implementing a software architecture using pre-existing components. Using object wrappers, the legacy code is encapsulated in a controllable way. The wrapper implements the interface through which newly developed objects access legacy resources; it is in charge of mapping incoming messages onto calls to the programs that implement the object methods. In the literature, different approaches for wrapping have been presented. Also, Mowbray and Zahari (1994) discuss different forms of wrapping (layering, data migration, middleware, encapsulation, wrappers for architecture implementation, wrappers for mediator and brokers), and present many alternative modes to implement a wrapper (remote procedure calls, files, sockets, application program interface, events, shared memory, macros).

Arranga and Coyle (1996) discuss the wrapping of legacy COBOL programs and present two methods: wrapping the entire program into a single service, and partitioning a program into individual services. The first method is possible only if the program can be reused as a whole. In that case a method is written to invoke the program using the adequate parameters. In particular, after the user interface component has been separated from the original program, the server part is reengineered to a multi-entry program; entry points are represented by any of the subroutines called by the user-interface component (Sneed, 1996). The second method requires splitting the old program into independently executable programs that can be invoked by different methods of the object wrapper. Canfora et al. (1999) exploit dominance analysis on the call graph to partition the subroutines into different programs and identify the entry subroutines of these programs. For example, with reference to Figure 6(b), the subroutines s5, s14, s15, s16, s17, and s18 are encapsulated in different programs; the subtrees rooted in s5 and s15 are also clustered in the programs obtained from these subroutines. Whatever wrapping strategy is used, data-flow analysis is necessary to identify the parameters exchanged by the user interface and the server programs (Cimitile, DeCarlin, & De Lucia, 1998) and enable the integration of the two parts in the Web-based architecture.

Wrappers can be realized in different ways. Sneed (1996) implements wrappers using CORBA IDL specifications to encapsulate ASSEMBLER and COBOL programs at different levels, including job, transaction, program, module, and procedure levels. Canfora et al. (1999) implement wrappers for RPG programs by exploiting the facilities provided by Visual Age C++ for AS/400. This allows RPG record formats to be mapped onto C structures and C functions to be associated with external programs. The Web offers several other powerful mechanisms for wrapping legacy systems in a distributed environment (Zou & Kontogiannis, 1999). The XML is an example of these technologies. It is widely used by developers as a standard format for data representation of the component's interfaces in a component repository, message exchanging, and for defining scripting languages (Sneed, 2000).

EXPERIENCE

The Web migration strategy outlined in the previous section has been applied in a pilot project concerned with a COBOL system, named Overmillion, produced by a small Italian software house. Overmillion is an online analytical processing system aimed at querying large archives of data for decision support through statistical and explorative analyses. It uses a proprietary form of inverted file, named FOPA (attribute-based file organization), which allows achieving very short response times (usually units of seconds also for archives with million of records) for this kind of analyses.

The system has evolved over the past 15 years from a centralized mainframe version with a character-based user interface to a client-server version with a graphical user interface running on a PC with the Microsoft Windows operating system. Only the user interface is separated from the rest of the system, while the application logic and the database services (server part) are not separated. Besides the original mainframe version based on VSAM files, the server part has been specialized to different platforms, such as IBM AS/400 and Aix, and different database management systems, such as the native AS/400 database and the DB2 relational database. Batch utilities convert the archive/database content to the FOPA format.

In the current version, the system provides the final user with two alternative work modalities: PC stand alone and remote host mode. Whenever the PC stand-alone mode is selected, the user works on a local image of the FOPA archive. In the remote mode, the user interface interacts with a set of COBOL programs distributed between the client and the server, which implements a socket-based connection with the remote host. The code implementing the querying and analyses of the FOPA archives is duplicated in different versions for the PC and the different hosts.

Several reasons motivated the need to migrate Overmillion to the Web. The first and most important is the fact that the system is used by several banks that use it to make decisions both at central and peripheral levels. In the old architecture, the PC-host connection layer had to be replicated on each client installation, and this increased the application ownership costs. Most banks nowadays use Internet/ Intranet technologies for their information system infrastructure, and this caused a push in the direction of Web migration. In addition, the new architecture adds flexibility; in particular, it opens the way to the migration of the stand-alone PC version towards a distributed version that federates several archives/databases host sources; the querying and analysis software based on FOPA organization will be

Figure 7: Architecture of the legacy system.

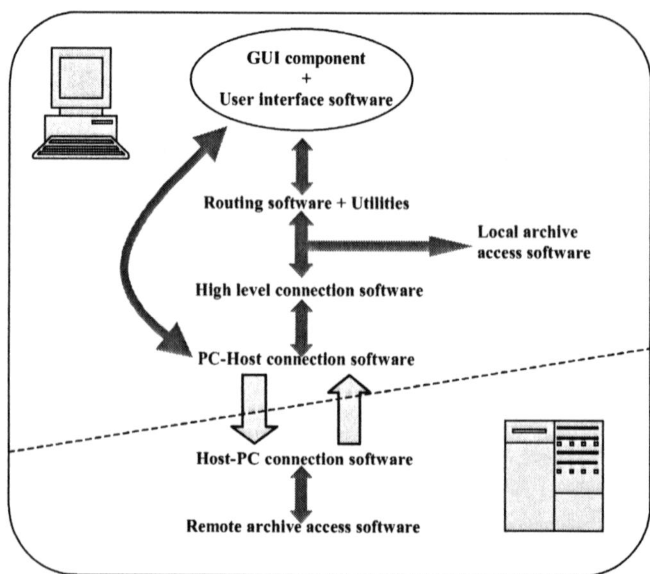

only on the middle tier, while the remote hosts will act as database servers for production transactions. This will eliminate the need to maintain the different versions of the querying and analysis software based on FOPA archives.

Assessing and Rearchitecting the System.

According to the classification by Brodie and Stonebraker (1995), the Overmillion system can be considered at least semidecomposable. Therefore, it can be collocated in the upper-right quadrant of the framework in Figure 3. Figure 7 shows the overall architecture of the system with a particular emphasis on the software running on the PC. The architecture was not formally recorded; we have reverse engineered it through code analysis and interviews with maintenance personnel and users of the system. In particular, we have used a commercial tool to extract the call graph of the system and a utility developed in house to transform it into a hierarchy by applying dominance analysis (Cimitile & Visaggio, 1995). Interviews have been used to confirm or correct the findings of code analysis.

Besides the graphical user interface component, the rest of the software running on the PC is written in COBOL. The graphical user interface is generated through a GUI generator and exploits a set of COBOL programs to make simple computations related to the management of the interaction objects (e.g., size, position). The graphical user interface and this set of programs implement the client-side of the system. The client communicates with a layer of COBOL utility programs and routing programs. The former programs implement general utility services, such as listing the registered archives and archives' fields, and exporting/importing archives of different formats; the latter provide the access to the stand-alone, local version of the system or the connection to the remote host, depending on the selected work mode. The programs that implement the connection to the remote host are split in two layers; in some cases the user-interface component directly communicates with the lower PC-Host connection software layer.

The migration project has focused on the software running on the PC; indeed, due to the particular architecture of the system, the software running on the remote

Table 1: Classification of PC COBOL programs.

Software Layer	# Programs	KLOC
User Interface	118	60
Routing and Utilities	28	11
Local Archive Access	31	41
High Level Connection	17	15
PC-Host Connection	8	5
Low Level Services	16	5

host does not need to be modified to enable the migration of the system to the Web. Besides the graphical user interface component, the migrated part of the system consisted of 218 COBOL programs and more than 130 KLOC. Table 1 shows a summary of the analyzed programs, classified according to the software layers of the architecture depicted in Figure 7. The low-level service software layer (last row in the table) consists of a set of programs that implement functions used by programs of other layers, in particular the user interface and the local archive access layers.

The particular architecture of the system (semidecomposable) did not require a particular effort to decompose it. Call graph restructuring was reduced to a minimum, as very few server programs issued calls to programs belonging to the user interface. More extensive restructuring activities were performed to transform global data areas used for the communication between the user interface and the other programs into parameters exchanged through LINKAGE SECTIONs; data flow analysis techniques (Cimitile et al., 1998) were extensively used to implement this task.

One of the aims of our pilot project was to achieve a balance between the effort required to reengineer or redevelop the legacy programs and the effort required to wrap them. For example, most of the programs that implemented the user interface component were very small (on the average 500 LOC including the DATA DIVISION); therefore, it was more convenient to redevelop this software layer using the same Web technology adopted for reengineering the user interface. With this approach, the minimization of the wrapping effort was achieved by wrapping only the programs of the Routing and Utility layer and the PC-Host lower level connection layer. In particular, the second group of programs was wrapped to

Figure 8: Web-enabled architecture of system.

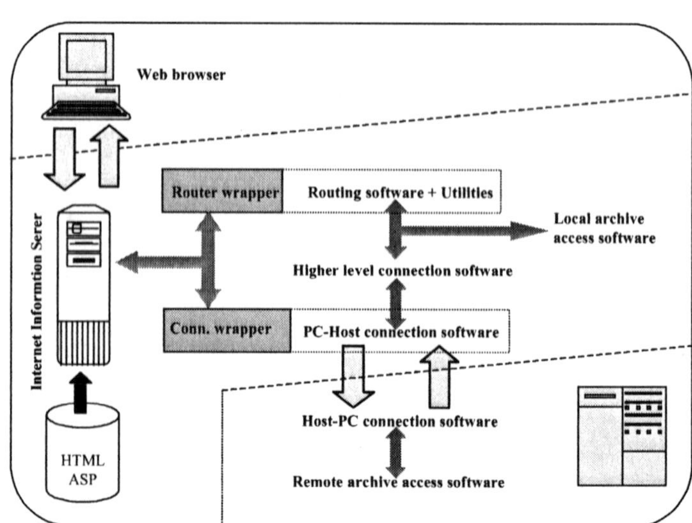

enable the communication with the new graphical user interface, while the interactions with the programs of the high communication layer is still implemented through the LINKAGE SECTION of the COBOL programs.

The target architecture of the system is shown in Figure 8. The two object wrappers, one for the routing software and the utilities and the other for the lower level connection layer, are realized as dynamic load libraries written in Microfocus Object COBOL; they are loaded into the Microsoft Internet Information Server (IIS) and are accessed by the user interface through VBScript functions embedded into ASP pages.

User Interface Reengineering, Wrapping, and Integration

In our project the technologies used to reengineering/redeveloping the user interfaces were ASP and VBScript. ASP is a flexible solution to quickly develop interactive Web pages using executable scripts in HTML pages. The scripting code is executed on the Web server before the page is sent to the browser; in this way, the Web application can work with any browser. The choice of ASP was mainly a consequence of the fact that the software company involved in the project was committed to Microsoft IIS as its platform for the development of Web applications and infrastructures. In addition, the company's personnel had previous experiences using Visual Basic (not VBScript) to develop traditional, two-tier client-server applications. Finally, the development workbench (Microsoft Visual Interdev) appeared to be rather powerful and relatively easy to learn.

To map the graphical objects of the old user interface onto objects of the new Web-based interface, we used the guidelines of the MORPH methodology (Moore, 1998). In our case, the reengineering of the Overmillion user interface was conducted manually; nevertheless, MORPH guidelines were useful to build a mapping between the interaction objects of the legacy user interface and the HTML objects of the new interface. The transformation of the legacy user interface into the new HTML interface was driven by the need to avoid retraining the old users of the system. Therefore, in addition to establishing a one-to-one correspondence between the legacy and the new interaction objects, the mapping we built maintained a correspondence between the legacy panels and the new HTML pages and forms. Details are provided in Aversano et al. (2001a).

To enable the communication between the new Web user interface and the COBOL legacy programs, two object wrappers have been developed—one for the routing software and the utilities and the other for the lower level connection layer (see Figure 8). The two wrappers have been realized as dynamic load libraries written in Microfocus Object COBOL and loaded into the IIS Web server. The wrapper interfaces provide a method for each legacy program to be accessed. For example, for the function *Somma,* the wrapper provides the method shown in

Figure 9. The wrapper receives messages from the user interface through the VBScript functions embedded into the ASP pages. The messages received by the wrapper are converted into calls to the programs that realize the required services.

The communications between the user interface and the wrapper is realized through strings. The VBScript functions build an input string instring from the data contained in the HTML form and pass them as a parameter to the object wrapper method (in the formal parameter VarIn); the string instring also contains information needed to identify the user and the session that issued the call. Figure 10 shows the fragment of VBScript function that invokes the method MSomma. The sequence of data concatenated in the input and output strings are mapped onto the structure of records of the invoked COBOL method.

The invoked method first extracts the data from the string VarIn into the fields of the local storage record APP-INFO2; then, relevant data is extracted from the fields of record APP-INFO2 and copied into the actual parameters to be passed to the legacy program. Finally, the wrapper method concatenates the parameters returned by the program into the string VarOut and returns it to the invoking scripting function. Special characters are used to separate record fields in the output string with a convention that uses different characters to identify the different nesting levels of the record fields (in the input string this is not required, as the concatenated fields have a fixed size). For example, the character "#" is used to separate fields of the outermost nesting level, while "=" is used for the second level of nesting. The VBScript code repeatedly splits the data contained in the returned string from the outermost to the innermost nesting level (see Figure 10) and produces the output for the appropriate client.

Figure 9: An excerpt from the object wrappers.

```
method-id. "MSomma".
local-storage Section.
01 APP-INFO2.
......
linkage Section.
01 VarIn        PIC X(1920).
01 VarOut        PIC X(1920).
......
procedure division using by reference VarIn
           returning VarOut.
....extracting parameters from string VarIn
  CALL "C:\OMWeb\Bin\SCSOMMA" USING .....
....concatenating parameters into string VarOut
```

CONCLUSION

Current business processes have been profoundly fitted to the available hardware and software. Traditionally, software systems have been developed with a mainframe-oriented, centralized architecture, and this has impacted the way businesses were conceived and conducted. Nowadays, the convergence between telecommunications and computing and the explosion of the Internet enables new ways of interaction between an organization and its customers and partners. As a consequence, moving to Web-centric, open architectures is widely recognized as a key to staying competitive in the dynamic business world. However, Internet and the Web are only enabling factors. To take advantage of the Internet's open architecture, most companies are applying business reengineering with the aim of moving from hierarchical centralized structures to networked decentralized business units cooperating with one another. Abstractly, reengineering business processes should entail discarding the existing and legacy systems to develop new software systems that meet the new business needs and have a distributed, Web-centric, component-based, open architecture. In practice, discarding existing and legacy systems is, in most cases, infeasible because they represent economic and knowledge investments to be salvaged, and the risks and costs associated with replacement cannot be afforded. As a consequence, the migration of the legacy systems to the new business and technological platform is often the only viable alternative. This means that a balance must be struck between the constraints imposed by the legacy systems and the opportunities offered by BPR.

Figure 10: An example of VBScript code for wrapper invocation.

```
......
result = Split (dataSource.MSomma (instring),"#")
Response.Write("<BR>Soggetti contati...............” & Cstr (CLng (result (0))))
Response.Write("<BR>Soggetti dell'applicazione.....” & Cstr (CLng (result (1))))
Response.Write("<BR>Data ultimo caricamento........” & result (2))
Response.Write("<BR>Tempo di elaborazione..........” & result (3))
Response.Write("<BR><BR><HR>")
for i = 4 to ubound (result) - 2
    res2 = Split (result(i),"=")
    s = sepDigit (Cstr (CDbl (res2(0))))
    m = sepDigit (Cstr (CDbl (res2(1))))
    if (fil(i - 4) <> "") then
     Response.Write("<BR>" & Session ("nameArray")(Cint(fil(i - 4))) & _
            "<BR>SOMMA = " & s & "<BR>MEDIA = " & m & "<HR>")
    end if
next
......
```

This chapter has proposed a strategy for migrating business processes and the supporting legacy systems toward an open, networked, Web-centric architecture. The initial step consists of modelling the existing business processes and assessing the business and technical value of the supporting software systems. The analysis of the existing processes is required to get an inventory of the activity performed, compare them with best practices, and redesign and/or reengineer them. The assessment of the legacy systems aims at evaluating its technical and business quality, to devise the most appropriate evolution approach. Based on the assessment, a decisional framework assists software managers to make informed decisions. This is followed by the migration of the legacy system, which can in turn be enacted with different approaches. The chapter proposed a short-term system migration strategy that decomposes and reengineers the system into its client and server components and uses wrappers to encapsulate the server components. Reverse engineering is used to abstract a model of the client components and, in particular, the user interface, and redevelop them. Enabling technologies for each of the activities involved in the migration process have been overviewed in the chapter.

Finally, the chapter has discussed the application of the overall migration strategy in an industrial pilot project. The project aimed at integrating an existing COBOL system into a Web-enabled architecture. The need for migrating the system was imposed by the changes in the underlying business processes. After assessing the system, we opted for a short-term migration strategy with the aim of reducing to a minimum the time needed to have a working version of the new, Web-enabled system and the extent of the changes to be made to the existing code. This was achieved essentially through an extensive use of reverse engineering to design the new user-interface, and wrapping to implement the server side. In particular, the presentation layer of the original system was separated from the application logic and database components and reimplemented using ASP and VBScript; the MORPH approach (Moore, 1998) was used to map the components of the existing interface onto the new Web-based interface. The application logic and database functions of the legacy system were wrapped into a single server component using dynamic load libraries written in Microfocus Object COBOL and accessed through VBScript functions embedded into ASP pages.

The migration project consumed approximately eight man/months over a period of five calendar months. Overall, five people were involved in the project: two software engineers from the company that had developed the original system, 2 junior researchers from the University, and a senior researcher with the role of coordination and advising. Assessing the existing system and decomposing it was performed jointly by an engineer of the company (he had also been involved in the development of the original system) and a junior researcher. Another joint team was in charge of selecting the enabling technologies, based on an assessment of the company's expertise and capabilities, and designing the new overall architecture of the system. At first, the development of the server and the client proceeded independently, with the company's engineers working on the COBOL wrapper and the university junior researchers working on the ASP integration layer and the client, including the reverse engineering of the original user interface. Server stubs were used to incrementally test the client and, as new features were integrated into the client, to test the corresponding server functions. This strategy of independent development was chosen because team members were expert in either COBOL or HTML and ASP, but not both. However, the last six calendar weeks of the project were spent on system testing, including fine-tuning the integration of client and server, and required a close cooperation of all the members of the team.

While the short-term strategy was successful in reducing the time and the costs of migrating the system to adapt it to the new business scenario, it did not changed the quality of the running server programs, which remain essentially unchanged, and therefore, no benefits are expected in the future maintenance and evolution of the system. This is in contrast with other approaches that reverse engineer meaningful objects from the legacy code and wrap each of them separately (see upper-left quadrant in Figure 3). While more costly, such approaches offer more long term benefits as they enable incremental replacement strategies.

REFERENCES

Arranga, E., & Coyle, F. (1996). *Object oriented COBOL*. New York: SIGS Books.

Aversano, L., Canfora, G., De Lucia, A., & Gallucci, P. (2002). Business process reengineering and workflow automation: A technology transfer experience. *The Journal of Systems and Software*, to appear.

Aversano, L., Canfora, G., & Stefanucci, S. (2001b). Understanding and improving the maintenance process: A method and two case studies. In *Proceedings of the 9th International Workshop on Program Comprehension*, (pp. 199-208) Toronto, Canada. New York: IEEE Computer Society.

Aversano, L., Cimitile, A., Canfora, G., & De Lucia, A. (2001a). Migrating legacy systems to the Web. In *Proceedings of the Conference on Software Maintenance and Reengineering*, (pp. 148-157) Lisbon, Portugal. New York: IEEE Computer Society.

Aversano, L., De Lucia, A., & Stefanucci, S. (2001c). Introducing worfklow management in software maintenance processes. In *Proceedings of the International Conference on Software Maintenance*, (pp. 441-450) Florence, Italy. New York: IEEE Computer Society.

Bennett K. H. (1995). Guest Editor. *IEEE Software Special Issue on Legacy Systems, 12*(1).

Bennett, K. H., Ramage, M., & Munro, M. (1999). Decision model for legacy systems. *IEE Proceedings Software, 146*(3), 153-159.

Booch, G., Rumbaugh, J., & Jacobson, I. (1999). *The Unified Modelling Language User Guide.* Reading, MA: Addison-Wesley.

Brodie, M.L., & Stonebaker, M. (1995). *Migrating Legacy Systems —Gateways, Interfaces & Incremental Approach.* San Francisco, CA: Morgan Kaufmann.

Canfora, G., & Cimitile, A. (1997). A reference life-cycle for legacy systems. In *Proceedings of the ICSE-19 Workshop on Migration Strategies for Legacy Systems.* Boston, MA, Technical Report TUV-1841-97-06, Technical University of Vienna, Information Systems Institute.

Canfora, G., Cimitile, A., De Lucia, A., & Di Lucca, G.A. (2000). Decomposing legacy programs: A first step towards migrating to client-server platforms. *The Journal of Systems and Software, 54*(2), 99-110.

Canfora, G., Cimitile, A., De Lucia, A. & Di Lucca, G.A. (2001). Decomposing legacy systems into objects: An eclectic approach. *Information and Software Technology, 43*(6), 401-412.

Canfora, G., Cimitile, A., & Munro, M. (1996). An improved algorithm for identifying reusable objects in code. *Software—Practice and Experiences,* 26, 24-48.

Canfora, G., De Lucia, A., & Di Lucca, G.A. (1999). An incremental object-oriented migration strategy for RPG legacy systems. *International Journal of Software Engineering and Knowledge Engineering, 9*(1), 5-25.

Cantone, G. (2000). Measure-driven processes and architecture for the empirical evaluation of software technology. *Journal of Software Maintenance: Research and Practice, 12*(1), 47-78.

Casati, F., Ceri, S., Pernici, B., & Pozzi, G., (1995). Conceptual modeling of workflows. *Proceedings of the 14th Object-Oriented and Entity-Relationship Modelling International Conference,* 341-354.

Cimitile, A., & Visaggio, G. (1995). Software salvaging and the call dominance tree. *The Journal of Systems and Software, 28*(2), 117-127.

Cimitile, A., De Carlini, U., & De Lucia, A. (1998). Incremental migration strategies: Data flow analysis for wrapping. *Proceedings of 5th Working Conference on Reverse Engineering,* (pp. 59-68) Honolulu, Hawaii: IEEE Computer Society.

Cimitile, A., De Lucia, A., Di Lucca, G.A., & Fasolino, A.R. (1999). Identifying objects in legacy systems using design metrics. *The Journal of Systems and Software, 44*(3), 199-211.

De Lucia, A., Fasolino, A.R., & Pompella, E. (2001). A decisional framework for legacy system management. *Proceedings of the International Conference on Software Maintenance*, (pp. 642-651) Florence, Italy: IEEE Computer Society.

Dietrich, W.C., Nackman, L.R., & Gracer, F. (1989). Saving a legacy with objects. *Proceedings of the Conference on Object Oriented Programming Systems Languages and Applications*, 77-88.

Dunn, M.F., & Knight, J.C. (1993). Automating the detection of reusable parts in existing software. In *Proceedings of 15th International Conference on Software Engineering*, (pp. 381-390) Baltimore, MD: IEEE Computer Society.

Hammer, M., & Champy, J. (1993). *Reengineering the Corporation: A Manifesto for Business Revolution*, New York: HarperCollins.

Jacobson, I., Ericsson, M., & Jacobson, A. (1995). *The Object Advantage: Business Process Reengineering with Object Technology*, ACM Press, Reading, MA: Addison-Wesley.

Kim, H.S., & Kwon, Y.R. (1994). Restructuring programs through program slicing. *International Journal of Software Engineering and Knowledge Engineering*, *4*(3), 349-368.

Lanubile, F., & Visaggio, G. (1997). Extracting reusable functions by flow graph-based program slicing. *IEEE Transactions on Software Engineering*, *23*(4), 246-259.

Lindig, C., & Snelting, G. (1997). Assessing modular structure of legacy code based on mathematical concept analysis. *Proceedings of 19th International Conference on Software Engineering*, (pp. 349-359) Boston, MA: ACM Press.

Liu, S., & Wilde, N. (1990). Identifying objects in a conventional procedural language: An example of data design recovery. *Proceedings of International Conference on Software Maintenance*, (pp. 266-271) San Diego, CA, IEEE Computer Society.

Livadas, P.E., & Johnson, T. (1994). A new approach to finding objects in programs. *Journal of Software Maintenance: Research and Practice*, 6, 249-260.

Loops, P., & Allweyer, T. (1998). Object orientation in business process modelling through applying event driven process chains (EPC) in UML. *Proceedings of 2nd International Workshop on Enterprise Distributed Object Computing*, (pp. 102-112) San Diego, CA: IEEE Computer Society.

Markosian, L., Newcomb, P., Brand, R., Burson, S., & Kitzmiller, T. (1994).

Using an enabling technology to reengineer legacy systems. *Communications of the ACM, 37*(5), 58-70.

Merlo, E., Gagne, P.Y., Girard, J.F., Kontogiannis, K., Hendren, L., Panangaden, P., & De Mori, R. (1995). Reengineering user interfaces. *IEEE Software, 12*(1), 64-73.

Montes de Oca, C., & Carver, D. (1998). Identification of data cohesive subsystems using data mining techniques. *Proceedings of International Conference on Software Maintenance*, (pp. 16-23) Bethesda, MD: IEEE Computer Society.

Moore, M. (1998). *User Interface Reengineering.* PhD Dissertation, College of Computing, Georgia Institute of Technology, Atlanta, GA.

Moore, M., & Moshkina, L. (2000). Migrating legacy user interfaces to the Internet: shifting dialogue initiative. *Proceedings of 7th Working Conference on Reverse Engineering*, (pp. 52-58) Brisbane, Australia: IEEE Computer Society.

Moore, M., & Rugaber, S. (1997). Using knowledge representation to understand interactive systems. *Proceedings of 5th International Workshop on Program Comprehension*, (pp. 60-67) Dearborn, MI.

Mowbray, T., & Zahari, R. (1994). *The Essential CORBA.* New York: John Wiley & Sons.

Müller, H.A., Orgun, M.A., Tilley, S.R., Uhl, J.S. (1993). A reverse-engineering approach to subsystem structure identification. *Journal of Software Maintenance: Research and Practice*, 5, 181-204.

Pigoski, T.M. (1997). *Practical Software Maintenance—Best Practices for Managing Your Software Investment.* New York: John Wiley & Sons.

Salminen, A., Lytikainen, K., & Tiitinen, P. (2000). Putting documents into their work context in document analysis. *Information Processing and Management, 36*(4), 623-642.

Serrano, M.A., Montes de Oca, C., & Carter, D.L. (1999). Evolutionary migration of legacy systems to an object-based distributed environment. *Proceedings of International Conference on Software Maintenance*, (pp. 86-95) Oxford, UK: IEEE Computer Society.

Siff, M., & Reps, T. (1999). Identifying modules via concept analysis. *IEEE Transactions on Software Engineering, 25*(6), 749-768.

Sneed, H.M. (1995). Planning the reengineering of legacy systems. *IEEE Software, 12*(1), 24-34.

Sneed, H.M. (1996). Encapsulating legacy software for use in client/server systems. *Proceedings of 3rd Working Conference on Reverse Engineering*, (pp. 104-199) Monterey, CA: IEEE Computer Society.

Sneed, H.M. (1999). Risks involved in reengineering projects. *Proceedings of 6th*

Working Conference on Reverse Engineering, (pp. 204-211) Atlanta, GA: IEEE Computer Society.

Sneed, H.M. (2000). Business reengineering in the age of the Internet. Technical Report, Case Consult, Wiesbaden, Germany.

Sneed, H.M., & Nyary, E. (1994). Downsizing large application programs. *Journal of Software Maintenance: Research and Practice*, *6*(5), 105-116.

Van der Aalst, W.M.P., (1998). The application of petri nets to workflow management. *The Journal of Circuits, Systems and Computers*, *8*(1), 21-66.

Van Sickle, L., Liu, Z.Y., & Ballantyne, M. (1993). Recovering user interface specifications for porting transaction processing applications. *Proceedings of 2nd Workshop on Program Comprehension*, (pp. 71-76) Capri, Italy: IEEE Computer Society.

Verdugo, G. (1988). Portfolio analysis - managing software as an asset. *Proceedings of International Conference Software Maintenance Management*, (pp. 17-24) New York.

Wiggerts, T.A. (1997). Using clustering algorithms in legacy system remodularization. *Proceedings of 4th Working Conference on Reverse Engineering*, (pp. 33-43) Amsterdam, The Netherlands: IEEE Computer Society.

Winograd, T., & Flores, F., (1986). *Understanding Computers and Cognition: A New Foundation for Design*. Norwood, NJ: Ablex Publishing.

Workflow Management Coalition. 1994. *Workflow Management Coalition: Reference Model*, Doc. no. WFMC-TC-1011, Brussels, Belgium. Available at http://www.wfmc.org/standards/docs/tc003v11.pdf.

Wu, B., Lawless, D., Bisbal, J., Wade, V., Grimson, J., Richardson, R., & O'Sullivan, D. (1997). The butterfly methodology: A gateway-free approach for migrating legacy information systems. *Proceedings of 3rd IEEE International Conference on Engineering of Complex Computer Systems,* (pp. 200-205) Como, Italy: IEEE Computer Society.

Zou, Y., & Kontogiannis, K. (1999). Enabling technologies for web based legacy system integration. *Proceedings of 1st International Workshop on Web Site Evolution*, (pp. 53-57) Atlanta, GA.

Chapter VII

Requirements Risk and Maintainability

Norman F. Schneidewind
Naval Postgraduate School, USA

In order to continue to make progress in software measurement as it pertains to reliability and maintainability, we must shift the emphasis from design and code metrics to metrics that characterize the risk of making requirements changes. By doing so we can improve the quality of delivered software, because defects related to problems in requirements specifications will be identified early in the life cycle. We provide an approach for identifying requirements change risk factors as predictors of reliability and maintainability problems. Our case example consists of twenty-four Space Shuttle change requests, nineteen risk factors, and the associated failures and software metrics. The approach can be generalized to other applications with numerical results that would vary according to application.

INTRODUCTION

While software design and code metrics have enjoyed some success as predictors of software quality attributes such as reliability and maintainability (Khoshgoftaar & Allen, 1998; Khoshgoftaar, Allen, Halstead, & Trio, 1996a; Khoshgoftaar, Allen, Kalaichelvan, & Goel, 1996b; Lanning & Khoshgoftaar, 1995; Munson & Werries, 1996; Ohlsson & Wohlin, 1998; Ohlsson & Alberg, 1996), the measurement field is stuck at this level of achievement. If measurement is to advance to a higher level, we must shift our attention to the front-end of the development process, because it is during system conceptualization that errors in specifying requirements are inserted into the process and adversely affect our ability to develop and maintain the software. A requirements change may induce ambiguity and uncertainty in the development process that cause errors in implementing the changes. Subsequently, these errors propagate through later phases of development and maintenance. These errors may result in significant risks associated with implementing the requirements. For example, reliability risk (i.e., risk of faults and failures induced by changes in requirements) may be incurred by deficiencies in the process (e.g., lack of precision in requirements). Although requirements may be specified correctly in terms of meeting user expectations, there could be significant risks associated with their implementation. For example, correctly implementing user requirements could lead to excessive system size and complexity with adverse effects on reliability and maintainability or there could be a demand for project resources that exceeds the available funds, time, and personnel skills. Interestingly, there has been considerable discussion of project risk (e.g., the consequences of cost overrun and schedule slippage) in the literature (Boehm, 1991) but not a corresponding attention to reliability and maintainability risk.

Risk in the Webster's New Universal Unabridged Dictionary (1979) is defined as "the chance of injury; damage, or loss." Some authors have extended the dictionary definition as follows: "Risk Exposure=Probability of an Unsatisfactory Outcome*Loss if the Outcome is Unsatisfactory" (Boehm, 1991). Such a definition is frequently applied to the risks in managing software projects such as budget and schedule slippage. In contrast, our application of the dictionary definition pertains to the risk of executing the software of a system where there is the chance of injury (e.g., crew injury or fatality), damage (e.g., destruction of the vehicle), or loss (e.g., loss of the mission) if a serious software failure occurs during a mission. We use risk factors to indicate the degree of risk associated with such an occurrence.

The generation of requirements is not a one-time activity. Indeed, changes to requirements can occur during maintenance. When new software is developed or existing software is changed in response to new and changed requirements, respectively, there is the potential to incur reliability and maintainability risks. Therefore, in assessing the effects of requirements on reliability and maintainability, we should deal with changes in requirements throughout the life cycle.

In addition to the relationship between requirements and reliability and maintainability, there are the intermediate relationships between requirements and software metrics (e.g., size, complexity) and between metrics and reliability and maintainability. These relationships may interact to put the reliability and maintainability of the software at risk because the requirements changes may result in increases in the size and complexity of the software that may adversely affect reliability and maintainability. We studied these interactions for the Space Shuttle. For example, assume that the number of iterations of a requirements change—the "mod level"—is inversely related to reliability and maintainability. That is, if many revisions of a requirement are necessary before it is approved, this is indicative of a requirement that is hard to understand and implement safely—a risk that directly affects reliability and maintainability. At the same time, this complex requirement will affect the size and complexity of the code that will, in turn, have deleterious effects on reliability and maintainability.

OBJECTIVES

Our overall objective is to identify the attributes of software requirements that cause the software to be unreliable and difficult to maintain. Furthermore, we seek to quantify the relationship between requirements risk and reliability and maintainability. If these attributes can be identified, then policies can be recommended to the software engineering community for recognizing these risks and avoiding or mitigating them during development and maintenance. The objective of these policy changes is to prevent the propagation of high-risk requirements through the various phases of software development and maintenance.

Given the lack of emphasis in measurement research on the critical role of requirements, we are motivated to discuss the following issues:

- What is the relationship between requirements attributes and reliability and maintainability? That is, are there requirements attributes that are strongly related to the occurrence of defects and failures in the software?
- What is the relationship between requirements attributes and software attributes like complexity and size? That is, are there requirements attributes that are strongly related to the complexity and size of software?
- Is it feasible to use requirements attributes as predictors of reliability and maintainability? That is, can static requirements change attributes like the size of the change be used to predict reliability in execution (e.g., failure occurrence) and the maintainability of the code?
- Are there requirements attributes that can discriminate between high and low reliability and maintainability, thus qualifying these attributes as predictors of reliability and maintainability?
- Which requirements attributes pose the greatest risk to reliability and maintainability?

An additional objective is to provide a framework that researchers and practitioners could use for the following: 1) to analyze the relationships among requirements changes, complexity, reliability, and maintainability, and 2) to assess and predict reliability and maintainability risk as a function of requirements changes.

METHODS

Our approach involves postulating several hypotheses concerning how requirements attributes affect reliability and maintainability and then conducting experiments to accept or reject the hypotheses. Various statistical methods can be used to identify the major risk factor contributors to unreliable and non-maintainable software. We illustrate selected methods using requirements and reliability data from the NASA Space Shuttle.

Several projects have demonstrated the validity and applicability of applying metrics to identify fault prone software at the code level (Khoshgoftaar & Allen, 1998; Khoshgoftaar et al., 1996a; Khoshgoftaar et al., 1996b; Schneidewind, 2000). Now, we apply this approach at the requirements level to allow for early detection of reliability and maintainability problems. Once high-risk areas of the software have been identified, they would be subject to detailed tracking throughout the development and maintenance process.

This chapter is organized as follows: background, selected measurement research projects, approach to analyzing requirements risk, risk factors, solutions to risk analysis example, future trends, and conclusions.

BACKGROUND

This topic is significant because the field of software engineering lacks the capability to quantitatively assess and predict the effect of a requirements change on the reliability and maintainability of the software. Much of the research and literature in software metrics concerns the measurement of code characteristics (Munson & Werries, 1996; Nikora, Schneidewind, & Munson, 1998). This is satisfactory for evaluating product quality and process effectiveness once the code is written. However, if organizations use measurement plans that are limited to measuring code, they will be deficient in the following ways: incomplete, lacking coverage (e.g., no requirements analysis and design), and starting too late in the process. For a measurement plan to be effective, it must start with requirements and continue through to operation and maintenance. Since requirements characteristics directly affect code characteristics and hence reliability and maintainability, it is important to assess their impact on reliability and maintainability when requirements are specified. We show that it is feasible to quantify the risks to reliability and maintainability of requirements changes—either new requirements or changes to existing requirements.

Once we are able to identify requirements attributes that portend high-risk for the operational reliability and maintainability of the software, it is then possible to suggest changes in the development and maintenance process of the organization. To illustrate, a possible recommendation is that any requirements change to mission critical software—either new requirements or changes to existing requirements—would be subjected to a *quantitative* risk analysis. In addition to stating that a risk analysis would be performed, the policy would specify the risk factors to be analyzed (e.g., number of modifications of a requirement or *mod level*) and their threshold or critical values. We have demonstrated the validity and applicability of identifying critical values of metrics to identify fault prone software at the code level (Schneidewind, 2000). For example, on the Space Shuttle, rigorous inspections of requirements, design documentation, and code have contributed more to achieving high reliability and maintainability than any other process factor. Thus, it would be prudent to consider adapting this process technology to other NASA projects, DoD, and other space and defense organizations because the potential payoff in increased reliability and maintainability would be significant. The objective of these policy changes is to prevent the propagation of high-risk requirements through the various phases of software development and maintenance. The payoff to these organizations would be to reduce the risk of mission critical software *not* meeting its reliability and maintainability goals during operation. For example, if the risk analysis identifies requirements that appear risky, measurements could be made on a prototype of the design and code to verify whether this is indeed the case. If the risk is confirmed through rapid prototyping, countermeasures could be considered such as modularizing or simplifying the requirements.

SELECTED MEASUREMENT RESEARCH PROJECTS

A number of reliability and maintenance measurement projects have been reported in the literature. For example, (Briand, Basili, & Kim, 1994) developed a process to characterize software maintenance projects. They present a qualitative and inductive methodology for performing objective project characterizations to identify maintenance problems and needs. This methodology aids in determining causal links between maintenance problems and flaws in the maintenance organization and process. Although the authors have related ineffective maintenance practices to organizational and process problems, they have not made a linkage to risk assessment.

Pearse and Oman (1995) applied a maintenance metrics index to measure the maintainability of C source code before and after maintenance activities. This technique allowed the project engineers to track the "health" of the code as it was being maintained. Maintainability is assessed but not in terms of risk assessment.

Pigoski and Nelson (1994) collected and analyzed metrics on size, trouble reports, change proposals, staffing, and trouble report and change proposal completion times. A major benefit of this project was the use of trends to identify the relationship between the productivity of the maintenance organization and staffing levels. Although productivity was addressed, risk assessment was not considered.

Sneed (1996) reengineered a client maintenance process to conform to the ANSI/IEEE Standard 1219, Standard for Software Maintenance. This project is a good example of how a standard can provide a basic framework for a process and can be tailored to the characteristics of the project environment. Although applying a standard is an appropriate element of a good process, risk assessment was not addressed.

Stark (1996) collected and analyzed metrics in the categories of customer satisfaction, cost, and schedule with the objective of focusing management's attention on improvement areas and tracking improvements over time. This approach aided management in deciding whether to include changes in the current release, with possible schedule slippage, or include the changes in the next release. However, the author did not relate these metrics to risk assessment.

An indication of the back seat that software risk assessment takes to hardware, Fragola (1996) reports on probabilistic risk management for the Space Shuttle. Interestingly, he says: "The shuttle risk is embodied in the performance of its hardware, the careful preparation activities that its ground support staff take between flights to ensure this performance during a flight, and the procedural and management constraints in place to control their activities." There is not a word in this statement or in his article about software! Another hardware-only risk assessment is by Maggio (1996), who says, "The current effort is the first integrated quantitative assessment of the risk of the loss of the shuttle vehicle from 3 seconds prior to lift-off to wheel-stop at mission end." Again, not a word about software. Pfleeger lays out a roadmap for assessing project risk that includes risk prioritization (Pfleeger, 1998), a step that we address with the degree of confidence in the statistical analysis of risk.

APPROACH TO ANALYZING REQUIREMENTS RISK

Our approach involves conducting experiments to see whether it is feasible to develop a mapping between changes in requirements to changes in software complexity and then to changes in reliability and maintainability. In other words, we investigate whether the following implications hold, where *R* represents requirements, *C* represents complexity, *F* represents failure occurrence (i.e., reliability), and *M* represents maintainability: $\Delta R \Rightarrow \Delta C \Rightarrow \Delta F, \Delta M$. We include changes in size and documentation in changes in complexity. We are able to judge whether the approach is a success by assessing whether this mapping can be achieved with the desired degree of statistical significance.

By retrospectively analyzing the relationship between requirements and reliability and maintainability, we are able to identify those risk factors that are associated with reliability and maintainability, and we are able to prioritize them based on the degree to which the relationship is statistically significant. In order to quantify the effect of a requirements change, we use various risk factors that are defined as the attribute of a requirement change that can induce adverse effects on reliability (e.g., failure incidence), maintainability (e.g., size and complexity of the code), and project management (e.g., personnel resources). Examples of Space Shuttle risk factors are shown in the RISK FACTORS section.

Table 1 shows the Change Request Hierarchy of the Space Shuttle, involving change requests (i.e., a request for a new requirement or modification of an existing requirement), discrepancy reports (i.e., reports that document deviations between specified and observed software behavior), and failures. We analyzed Categories 1 versus 2 with respect to risk factors as discriminants of the categories.

Categorical Data Analysis

Using the null hypothesis, Ho: A risk factor is not a discriminator of reliability and maintainability versus the alternate hypothesis H_1: A risk factor is a discriminator of reliability and maintainability, we used categorical data analysis to test the hypothesis. A similar hypothesis was used to assess whether risk factors can serve as discriminators of metrics characteristics. We used the requirements, requirements risk factors, reliability, and metrics data we have from the Space Shuttle *Three Engine Out* software (abort sequence invoked when three engines are lost) to test our hypotheses. Samples of these data are shown below.

- Pre-release and post release failure data from the Space Shuttle from 1983 to the present. An example of post-release failure data is shown in Table 2, where an Operational Increment (OI) is a software system comprised of modules and configured from a series of builds to meet Space Shuttle mission functional requirements.
- Risk factors for the Space Shuttle *Three Engine Out Auto Contingency* software. This software was released to NASA by the developer on October 18, 1995. An example of a partial set of risk factor data is shown in Table 3.

Table 1. Change request hierarchy

Change Requests (CRs)
1. No Discrepancy Reports (i.e., CRs with no DRs)
2. (Discrepancy Reports) or (Discrepancy Reports and Failures)
2.1 No failures (i.e., CRs with DRs only)
2.2 Failures (i.e., CRs with DRs and Failures)
2.2.1 Pre-release failures
2.2.2 Post-release failures

Table 2.

Failure Found On Operational Increment	Days from Release When Failure Occurred	Discrepancy Report #	Severity	Failure Date	Release Date	Module in Error
Q	75	1	2	05-19-97	03-05-97	10

Table 3.

Change Request Number	Source Lines of Code Changed	Complexity Rating of Change	Criticality of Change	Number of Principal Functions Affected	Number of Modifications of Change	Number of Requirements Issues	Number of Inspections Required	Manpower Required to Make Change
A	1933	4	3	27	7	238	12	209.3 Man Weeks

Table 4.

Module	Operator Count	Operand Count	Statement Count	Path Count	Cycle Count	Discrepancy Report Count	Change Request Count
10	3895	1957	606	998	4	14	16

- Metrics data for 1400 Space Shuttle modules, each with 26 metrics. An example of a partial set of metric data is shown in Table 4.

Table 5 shows the definition of the Change Request samples that were used in the analysis. Sample sizes are small due to the high reliability of the Space Shuttle. However, sample size is one of the parameters accounted for in the statistical tests that produced statistically significant results in certain cases (see the SOLUTIONS TO RISK ANALYSIS EXAMPLE).

To minimize the effects of a large number of variables that interact in some cases, a statistical categorical data analysis was performed incrementally. We used only one category of risk factor at a time to observe the effect of adding an additional risk factor on the ability to correctly classify change requests that have *No Discrepancy Reports* versus change requests that have ((*Discrepancy Reports Only*) *or* (*Discrepancy Reports and Failures*)). The Mann-Whitney Test for difference in medians between categories was used because no assumption need be made about statistical distribution. In addition, some risk factors are ordinal scale quantities (e.g., modification level); thus, the median is an appropriate statistic to

Table 5. Definition of samples

Sample	Size
Total CRs	24
CRs with no DRs	14
CRs with (DRs only) or (DRs and Failures)	10
CRs with modules that caused failures	6
CRs can have multiple DRs, failures, and modules that caused failures. CR: Change Request. DR: Discrepancy Report.	

use. Furthermore, because some risk factors are ordinal scale quantities, rank correlation (i.e., correlation coefficients are computed based on rank) was used to check for risk factor dependencies.

RISK FACTORS

One of the software process problems of the NASA Space Shuttle Flight Software organization is to evaluate the risk of implementing requirements changes. These changes can affect the reliability and maintainability of the software. To assess the risk of change, the software development contractor uses a number of risk factors, which are described below. The risk factors were identified by agreement between NASA and the development contractor based on assumptions about the risk involved in making changes to the software. This formal process is called a risk assessment. No requirements change is approved by the change control board without an accompanying risk assessment. During risk assessment, the development contractor will attempt to answer such questions as "Is this change highly complex relative to other software changes that have been made on the Space Shuttle?" If this were the case, a high-risk value would be assigned for the complexity criterion. To date, this *qualitative* risk assessment has proven useful for identifying possible risky requirements changes or, conversely, providing assurance that there are no unacceptable risks in making a change. However, there has been no quantitative evaluation to determine whether, for example, high-risk factor software was really less reliable and maintainable than low-risk factor software. In addition, there is no model for predicting the reliability of the software if the change is implemented. We address both of these issues.

We had considered using requirements attributes like completeness, consistency, correctness, etc., as risk factors (Davis, 1990). While these are useful generic concepts, they are difficult to quantify. Although some of the following risk factors also have qualitative values assigned, there are a number of quantitative risk factors, and many of the risk factors deal with the execution behavior of the software (i.e., reliability) and its maintainability, which is our primary interest.

SPACE SHUTTLE FLIGHT SOFTWARE REQUIREMENTS CHANGE RISK FACTORS

The following are the definitions of the nineteen risk factors, where we have placed the risk factors into categories and have provided our interpretation of the question the risk factor is designed to answer. If the answer to a yes/no question is "yes," it means this is a high-risk change with respect to the given risk factor. If the answer to a question that requires an estimate is an anomalous value, it means this is a high-risk change with respect to the given risk factor.

For each risk factor, it is indicated whether there is a statistically significant relationship between it and reliability and maintainability for the software version analyzed. The details of the findings are shown in the SOLUTIONS TO RISK ANALYSIS EXAMPLE section. In many instances, there was insufficient data to do the analysis. These cases are indicated below. The names of the risk factors used in the analysis are given in quotation marks.

Complexity Factors

- Qualitative assessment of complexity of change (e.g., very complex); "complexity." **Not significant**.
    - Is this change highly complex relative to other software changes that have been made on the Space Shuttle?
- Number of modifications or iterations on the proposed change; "mods." **Significant**.
    - How many times must the change be modified or presented to the Change Control Board (CCB) before it is approved?

Size Factors

- Number of source lines of code affected by the change; "sloc." **Significant**.
    - How many source lines of code must be changed to implement the change request?
- Number of modules changed; "mod chg." **Not significant**.
    - Is the number of changes to modules excessive?

Criticality of Change Factors

- Criticality of function added or changed by the change request; "crit func." (insufficient data)
    - Is the added or changed functionality critical to mission success?
- Whether the software change is on a nominal or off-nominal program path (i.e., exception condition); "off nom path." (insufficient data)

- Will a change to an off-nominal program path affect the reliability of the software?

Locality of Change Factors

- The area of the program affected (i.e., critical area such as code for a mission abort sequence); "critic area." (insufficient data)
  - Will the change affect an area of the code that is critical to mission success?
- Recent changes to the code in the area affected by the requirements change; "recent chgs." (insufficient data)
  - Will successive changes to the code in one area lead to non-maintainable code?
- New or existing code that is affected; "new\exist code." (insufficient data)
  - Will a change to new code (i.e., a change on top of a change) lead to non-maintainable code?
- Number of system or hardware failures that would have to occur before the code that implements the requirement would be executed; "fails ex code." (insufficient data)
  - Will the change be on a path where only a small number of system or hardware failures would have to occur before the changed code is executed?

Requirements Issues and Functions Factors

- Number and types of other requirements affected by the given requirement change (requirements issues); "other chgs." (insufficient data)
  - Are there other requirements that are going to be affected by this change? If so, these requirements will have to be resolved before implementing the given requirement.
- Number of possible conflicts among requirements (requirements issues); "issues." **Significant**.
  - Will this change conflict with other requirements changes (e.g., lead to conflicting operational scenarios)?
- Number of principal software functions affected by the change; "prin funcs." **Not significant**.
  - How many major software functions will have to be changed to make the given change?

Performance Factors

- Amount of memory space required to implement the change; "space." **Significant**.

  - Will the change use memory to the extent that other functions will not have sufficient memory to operate effectively?
- Effect on CPU performance; "cpu." (insufficient data)
  - Will the change use CPU cycles to the extent that other functions will not have sufficient CPU capacity to operate effectively?

Personnel Resources Factors

- Number of inspections required to approve the change; "inspects." **Not significant**.
  - Will the number of requirements inspections lead to excessive use of personnel resources?
- Manpower required to implement the change; "manpower." **Not significant**.
  - Will the manpower required to implement the software change be significant?
- Manpower required to verify and validate the correctness of the change; "cost." **Not significant**.
  - Will the manpower required to verify and validate the software change be significant?
- Number of tests required to verify and validate the correctness of the change; "tests." **Not significant**.
  - Will the number of tests required to verify and validate the software change be significant?

SOLUTIONS TO RISK ANALYSIS EXAMPLE

This section contains the solutions to the risk analysis example, using the Space Shuttle data and performing the statistical analyses in a. b., and c., as shown in Tables 6, 7, and 8, respectively. Only those risk factors where there was sufficient data and the results were statistically significant are shown.

a. Categorical data analysis on the relationship between (CRs with no DRs) versus ((DRs only) or (DRs and Failures)), using the Mann-Whitney Test;
b. Dependency check on risk factors, using rank correlation coefficients; and
c. Identification of modules that caused failures as a result of the CR, and their metric values.

Categorical Data Analysis

Of the original nineteen risk factors, only four survived as being statistically significant (alpha $\leq .05$); seven were not significant; and eight had insufficient data to make the analysis. As Table 6 shows, there are statistically significant results for (CRs with no DRs) versus ((DRs only) or (DRs and Failures)) for the risk factors "mods," "sloc," "issues," and "space." We use the value of alpha in Table 6 as a means to prioritize the use of risk factors, with low values meaning high priority. The priority order is: "issues," "space," "mods," and "sloc."

The significant risk factors would be used to predict reliability and maintainability problems for this set of data and this version of the software. Whether these results would hold for future versions of the software would be determined in validation tests on subsequent Operational Increments. The finding regarding "mods" does confirm the software developer's view that this is an important risk factor. This is the case because if there are many iterations of the change request, it implies that it is complex and difficult to understand. Therefore, the change is likely to lead to reliability and maintainability problems. It is not surprising that the size of the change "sloc" is significant because our previous studies of Space Shuttle metrics have shown it to be important determinant of software quality (Schneidewind, 2000). Conflicting requirements "issues" could result in reliability and maintainability problems when the change is implemented. The on-board computer memory required to implement the change "space" is critical to reliability and maintainability because unlike commercial systems, the Space Shuttle does not have the luxury of large physical memory, virtual memory, and disk memory to hold its programs and data. Any increased requirement on its small memory to implement a requirements change comes at the price of demands from competing functions.

In addition to identifying predictive risk factors, we must also identify thresholds for predicting when the number of failures would become excessive (i.e., rise rapidly with the risk factor). An example is shown in Figure 1, where cumulative failures are plotted against cumulative issues. The figure shows that when issues reach 286, failures reach 3 (obtained by querying the data point) and climb rapidly thereafter. Thus, an issues count of 286 would be the best estimate of the threshold to use in controlling the quality of the next version of the software. This process would be repeated across versions with the threshold being updated as more data is gathered. Thresholds would be identified for each risk factor in Table 6. This would

Table 6. Statistically significant results (alpha ≤ .05). CRs with no DRs vs. ((DRs only) or (DRs and Failures)). Mann-Whitney Test.

Risk Factor	**Alpha**	**Median Value** CRs with no DRs	**Median Value** (DRs only) or (DRs and Failures)
issues	.0076	1	14
space	.0186	6	123
mods	.0401	0	4
sloc	.0465	10	88.5
issues: Number of possible conflicts among requirements. space: Amount of memory space required to implement the change. mods: Number of modifications of the proposed change. Sloc: Number of source lines of code affected by the change.			

provide multiple alerts for the quality of the software going bad (i.e., the reliability and maintainability of the software would degrade as the number of alerts increases).

Dependency Check on Risk Factors

In order to check for possible dependencies among risk factors that could confound the results, rank correlation coefficients were computed in Table 7. Using an arbitrary threshold of .7, the results indicate a significant dependency among "issues," "mods," and "sloc" for CRs with no DRs. That is, as the number of

Table 7. Rank correlation coefficients of risk factors

CRs with no DRs				
	mods	sloc	issues	space
mods		.370	**.837**	.219
sloc	.370		**.717**	.210
issues	**.837**	**.717**		.026
space	.219	.210	.026	
(DRs only) or (DRs and Failures)				
	mods	sloc	issues	space
mods		.446	.363	**.759**
loc	.446		.602	.569
issues	.363	.602		**.931**
space	**.759**	.569	**.931**	

Figure 1. Failures vs. requirements issues

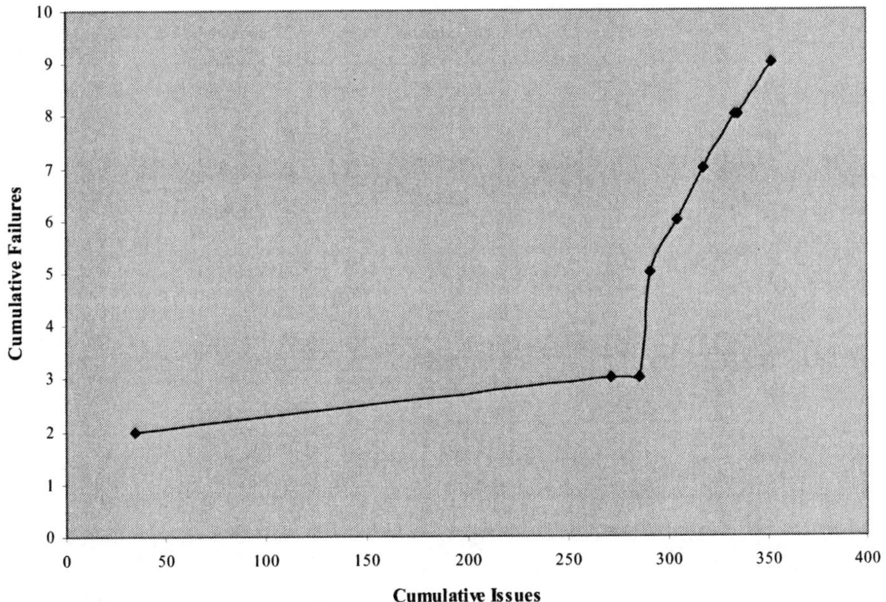

conflicting requirements increases, the number of modifications and size of the change increase. In addition, there is a significant dependency among "space," "mods," and "issues" for (DRs only) or (DRs and Failures). That is, as the number of conflicting requirements increases, the memory space and the number of modifications increase.

Identification of Modules that Caused Failures

Table 8 shows modules that caused failures as the result of the CRs had metric values that far exceed the critical values. The latter were computed in previous research (Schneidewind, 2000). A critical value is a discriminant that distinguishes high-quality from low-quality software. A module with metric values exceeding the critical values is predicted to cause failures and maintainability problems. Although the sample sizes are small, due to the high reliability of the Space Shuttle, the results consistently show that modules with excessive size and complexity lead to failures. Not only will the reliability be low but this software will also be difficult to maintain. The application of this information is that there is a high degree of risk when changes are made to software that has the metric characteristics shown in the table. Thus, these characteristics should be considered when making the risk analysis.

FUTURE TRENDS

Requirements risk analysis is another project in our series of software measurement projects that has included software reliability modeling and prediction, metrics analysis, and maintenance stability analysis (Schneidewind, 1998b; Schneidewind, 1997b). We have been involved in the development and application of software reliability models for many years (Schneidewind, 1993, 1997c; Schneidewind & Keller, 1992). Our models, as is the case in general in software

Table 8. Selected risk factor module characteristics

Change Request	Failed Module	Metric Critical Value	Metric Value	Metric
A	1	change history line count in module listing	63	558
A	2	non-commented loc count	29	408
B	3	executable statement count	27	419
C	4	unique operand count	45	83
D	5	unique operator count	9	33
E	6	node count (in control graph)	17	66
All of the above metrics exceeded the critical values for all of the above Change Requests.				

reliability, use failure data as the driver. This approach has the advantage of using a metric that represents the dynamic behavior of the software. However, this data is not available until the test phase. Predictions at this phase are useful but it would be much more useful to predict at an earlier phase—preferably during requirements analysis—when the cost of error correction is relatively low. Thus, there is great interest in the software reliability and metrics field in using static attributes of software in reliability and maintainability modeling and prediction. Presently, the software engineering field does not have the capability to make early predictions of reliability and maintainability problems. Early predictions would allow errors to be discovered and corrected when the cost of correction is relatively low. In addition, early detection would prevent poor quality software from getting into the hands of the user. As a future trend, the focus in research will be to identify the attributes of software requirements that cause the software to be unreliable and difficult to maintain.

Based on the premise that no one model suffices for all prediction applications, a future research goal is to develop an integrated suite of models for various applications. One type of model and its predictions have been described. Other members of our suite are quality metrics (Schneidewind, 1997a) and process stability models (Schneidewind, 1998a). We recommend that other researchers change their emphasis to the prediction of reliability and maintainability at the earliest possible time in the development process—to the requirements analysis phase—that heretofore has been unattainable. In doing so, researchers would have the opportunity to determine whether there exists a "standard" set of risk factors that could be applied in a variety of applications to reliability and maintainability prediction.

CONCLUSIONS

Our objective has been to improve the safety of software systems, particularly safety critical systems, by reducing the risk of failures attributed to software. By improving the reliability and maintainability of the software, where the reliability and maintainability measurements and predictions are directly related to safety, we contribute to improving system safety.

> *Risk factors that are statistically significant can be used to make decisions about the risk of making changes. These changes affect the reliability and maintainability of the software. Risk factors that are not statistically significant should not be used; they do not provide useful information for decision-making and cost money and time to collect and process. Statistically significant results were found for CRs with no DRs vs. ((DRs only) or (DRs and Failures)). The number of requirements issues ("issues"), the amount of memory*

space required to implement the change ("space"), the number of modifications ("mods"), and the size of the change ("sloc") were found to be significant for the Space Shuttle, in that priority order. In view of the dependencies among these risk factors, "issues" would be the choice if the using organization could only afford a single risk factor. We also showed how risk factor thresholds are determined for controlling the quality of the next version of the software.

Metric characteristics of modules should be considered when making the risk analysis because metric values that exceed the critical values are likely to result in unreliable and non-maintainable software.

Our methodology can be generalized to other risk assessment domains, but the specific risk factors, their numerical values, and statistical results may vary. Future research opportunities involve applying the methodology to the next version of the Space Shuttle software and identifying the statistically significant risk factors and thresholds to see whether they match the ones identified in this chapter. In addition, researchers may apply our methodology to other domains to test its validity on other applications.

ACKNOWLEDGMENTS

We acknowledge the technical support received from Julie Barnard, Boyce Reeves, and Patti Thornton of United Space Alliance. We also acknowledge the funding support received from Dr. Allen Nikora of the Jet Propulsion Laboratory.

REFERENCES

Boehm, W. (1991). Software risk management: Principles and practices. *IEEE Software*, 8 (1), 32-41.

Briand, L. C., Basili, V. R., & Kim, Y.-M. (1994). Change analysis process to characterize software maintenance projects. *Proceedings of the International Conference on Software Maintenance*, Victoria, British Columbia, Canada, 38-49.

Davis, A. (1990). *Software requirements: Analysis and specifications,* Englewood Cliffs, NJ: Prentice Hall.

Elbaum, S. G. & Munson, J. C. (1998). Getting a handle on the fault injection process: Validation of measurement tools. *Proceedings of the Fifth International Software Metrics Symposium*, Bethesda, MD 133-141.

Fragola, J. R. (1996). Space shuttle program risk management. *Proceedings of the Annual Reliability and Maintainability Symposium*, 133-142.

Khoshgoftaar, T. M., & Allen, E. B. (1998). Predicting the order of fault-prone modules in legacy software. *Proceedings of the Ninth International Symposium on Software Reliability Engineering*, Paderborn, Germany, 7, 344-353.

Khoshgoftaar, T. M., Allen, E. B., Halstead, R., & Trio, G. P. (1996a). Detection of fault-prone software modules during a spiral life cycle. *Proceedings of the International Conference on Software Maintenance*, Monterey, CA, 69-76.

Khoshgoftaar, T. M., Allen, E. B., Kalaichelvan, K, & Goel, N. (1996b). Early quality prediction: A case study in telecommunications. *IEEE Software*, 13 (1), 65-71.

Lanning, D. & Khoshgoftaar, T. (1995). The impact of software enhancement on software reliability. *IEEE Transactions on Reliability*, 44 (4), 677-682.

Maggio, G. (1996). Space shuttle probabilistic risk assessment methodology and application. *Proceedings of the Annual Reliability and Maintainability Symposium*, 121-132.

Munson, J. C. & Werries, D. S. (1996). Measuring software evolution. *Proceedings of the Third International Software Metrics Symposium*, Berlin, Germany, 41-51.

Nikora, A. P., Schneidewind, N. F., & Munson, J. C. (1998). IV&V Issues in achieving high reliability and safety in critical control software, final report, Vol. 1; Measuring and evaluating the software maintenance process and metrics-based software quality control, Vol. 2; Measuring defect insertion rates and risk of exposure to residual defects in evolving software systems, Vol. 3; and Appendices. Pasadena, CA Jet Propulsion Laboratory, National Aeronautics and Space Administration.

Ohlsson, M., C., & Wohlin, C., (1998). Identification of green, yellow, and red legacy components. *Proceedings of the International Conference on Software Maintenance,* Bethesda, MD, 6-15.

Ohlsson, N. & Alberg, H. (1996). Predicting fault-prone software modules in telephone switches. *IEEE Transactions on Software Engineering*, 22(12), 886-894.

Pearse, T. & Oman, P. (1995). Maintainability measurements on industrial source code maintenance activities. *Proceedings of the International Conference on Software Maintenance,* Opio (Nice), France, 295-303.

Pfleeger, S. L. (1998). Assessing project risk. *Software Tech N*ews, DoD Data Analysis Center for Software, 2(2), 5-8.

Pigoski, T. M. & Nelson, L. E. (1994). Software maintenance metrics: A case study. *Proceedings of the International Conference on Software Maintenance*, Victoria, British Columbia, Canada, 392-401.

Schneidewind, N. F. (1993). Software reliability model with optimal selection of failure data. *IEEE Transactions on Software Engineering*, 19(11), 1095-1104.

Schneidewind, N. F. (1997a). Software metrics model for integrating quality control and prediction. *Proceedings of the Eight International Symposium on Software Reliability Engineering*, Albuquerque, New Mexico, 402-415.

Schneidewind, N. F. (1997b). Measuring and evaluating maintenance process using reliability, risk, and test metrics. *Proceedings of the International Conference on Software Maintenance*, Bari, Italy, 232-239.

Schneidewind, N. F. (1997c). Reliability modeling for safety critical software. *IEEE Transactions on Reliability*, 46(1), 88-98.

Schneidewind, N. F. (1998a). An integrated process and product model. *Proceedings of the Fifth International Metrics Symposium*, Bethesda, Maryland, 224-234.

Schneidewind, N. F. (1998b). How to evaluate legacy system maintenance. *IEEE Software*, 15(4), 34-42. Also translated into Japanese and reprinted in: (pp 232-240). Tokyo, Japan: Nikkei Computer Books, Nikkei Business Publications, Inc.

Schneidewind, N. F. (2000). Software quality control and prediction model for maintenance. *Annals of Software Engineering*, 9, 79-101, Baltzer Science Publishers.

Schneidewind, N. F. & Keller, T. W. (1992). Application of reliability models to the space shuttle. *IEEE Software*, 9(4), 28-33.

Sneed, H. (1996). Modelling the maintenance process at Zurich Life Insurance. *Proceedings of the International Conference on Software Maintenance*, Monterey, CA, 217-226.

Stark, G. E. (1996). Measurements for managing software maintenance, *Proceedings of the International Conference on Software Maintenance*, Monterey, CA, 152-161.

Webster's New Universal Unabridged Dictionary (1979), Second Edition, New York: Simon & Schuster.

Chapter VIII

Software Maintenance Cost Estimation

Harry M. Sneed
Institut für Wirtschaftsinformatik, University of Regensburg, Bavaria

This chapter deals with the subject of estimating the costs of software maintenance. It reviews the existing literature on the subject and summarises the various approaches taken to estimate maintenance costs starting with the original COCOMO approach in 1981. It then deals with the subject of impact analysis and why it is essential to estimate the scope of maintenance projects. Examples are given to illustrate this. It then goes on to describe some of the tools the author has developed in the past ten years to support his practice of maintenance project estimation including the tools SoftCalc and MainCost. For both of these tools empirical studies of industrial experiments are presented as proof of the need to automate the estimation process.

THE RATIONALE FOR MAINTENANCE COST ESTIMATION

Software maintenance encompasses the tasks of correcting, adapting, enhancing, and perfecting a software product which is already in operation. (ANSI-IEEE, 1998) Except in the case of critical errors and unavoidable adaptations, the user has an option of whether to correct, adapt, enhance, or perfect the software, or not. The maintenance task is not imperative. Even in the case of non-critical errors, it is often possible to go on using the software as it is. The decision whether to carry through the maintenance task or not depends on two factors – time and cost.

The situation is similar to a person with a defective automobile which still functions. The owner of the automobile has the option of whether to fix it or not, as long as it is not a safety hazard. Whether he has it repaired or not depends on the time and costs. In any case, he will want to know what it will cost before he does anything. If he is prudent, he will visit several repair shops to get different estimates. If he happens to be in a foreign country and the repair will take a week, he may decide to wait until he returns home. Or, if the cost is too high, he may decide to wait until he has the money to pay for it.

The time and cost plays an even greater role in the case of enhancements and perfection. It is similar to the situation of a home owner who wants to either add a bathroom on to the house – enhancement – or renovate an existing bathroom – perfection. In neither case is he compelled to act now. He can postpone this task indefinitely depending on his budget. Everything is a question of costs and benefits and priorities in the light of the current economic situation.

Software system owners, like unfinished home owners, have a number of options of what they could do next. They could upgrade the user interface, add new system interfaces, add new functionality, or restructure the code. What actions they take depend on their current priorities and the costs of implementing the actions. There is always more that could be done, than what the owner has the capacity or money to have done. If the owner is rational, he will weigh the costs and benefits of each alternative action and select them on the basis of their cost/benefit relationship.

Therefore, except for critical errors and unavoidable adaptations, knowledge of the costs is essential. The owner of the system must know what a given maintenance task will cost in order to make a rational decision whether to pay for the task or not. It is up to the maintainers of the software to give him a cost and time offer, just as the car repair shop is obligated to make an offer for repair to the automobile owner or the plumber is obligated to make an offer for a new bath to the home owner. This is such a natural phenomenon that it is astounding to note how uncommon it occurs in the software business. One can only attribute the lack of a proper customer service relationship to the immaturity of the software field.

RESEARCH ON MAINTENANCE COST ESTIMATION

The lack of attention to a proper cost/benefit analysis of maintenance tasks is reflected in the pertinent literature. There has been very little research published on the subject, and what research there has been, has had little impact on industrial practice. One of the first attempts to estimate the costs of software maintenance is described by Barry Boehm in his book *Software Engineering Economics*. (1981). Boehm maintained that annual maintenance costs can be derived from the initial development effort – DevEffort– and the annual change traffic – ACT, adjusted by a multiplication factor for the system type – Type.

Boehm's basic maintenance effort equation computed maintenance effort as a function of development effort and system type:

Annual Maint Effort = Type (ACT) (DevEffort)

where

ACT = Annual Change Traffic, and

Type = Adjustment Factor from 1 to 3 based on type of system (batch, online, real time)

The extended COCOMO maintenance effort equation included quality and complexity as adjustment factors.

Annual Maint Effort = Annual Maint Effort (Comp ′ -Qual)

where

Comp = Complexity = Range (0,5 : 1,5), and

-Qual = Quality = –Range (0,5 : 1,5)

With these factors, the maintenance effort could be adjusted up or down depending on the quality and complexity of the product being maintained. Complexity increases effort whereas quality reduces it. It was the achievement of Boehm to bring complexity and quality to bear on the quantity of the software to be maintained (Boehm, 1983).

Boehm came to the conclusion in a study on maintenance costs in the U.S. Defense Department that maintenance costs are three to five times that of the original development costs, meaning all costs after the initial release of the systems. (Boehm, 1988). His statistics have been quoted by many others in trying to predict life cycle costs. Boehm's approach is, of course, useful at the macro planning level; however, it is of no use in trying to estimate the costs of individual maintenance tasks. Here, a micro planning approach is necessary.

Such an approach was presented by Card, Cotnoir and Goorevich (1987) at the 1987 Software Maintenance Conference. They suggested counting the number of lines of code affected by the maintenance action, then relating this count to effort, whereby they are careful to point out that productivity is much lower when making fixes and changes than when writing the code. For their model they used data acquired from maintenance projects at the Network Control Center of the Goddard Space Flight Center. The relationship between lines of code and effort was expressed in the equation:

SID = 42 + 4.8 NU ¸ 46.6 SPRF**

where

SID = person-days

NU = new code units to be implemented

SPRF = fixes and changes to be implemented.

Forty-two is the minimum number of days they claim is necessary to test any new release. The effort required to make fixes seems to be relatively high, but it is based on real data and has to be accepted. It is doubtful, however, if business data processing shops would be willing to pay so much for maintenance.

Another study which was directed toward assessing maintenance costs was that of Vessey and Weber (1993) in Australia. These authors studied 447 commercial COBOL programs to determine to what degree program complexity, programming style, programmer productivity, and the number of releases affects maintenance costs. Surprisingly, they came to the conclusion that program complexity has only a limited impact on repair costs. They also discovered that programming style is only significant in the case of larger programs. The number of releases only affected the costs of adaptive maintenance but not repair maintenance. The more a program is changed, the more difficult it is to adapt. This reinforced the conclusion that increasing complexity drives up adaptive maintenance costs. Thus, it is important to assess program complexity, whereas programming style, e.g., structuredness and modularity, is not so relevant as many would believe.

Dieter Rombach (1987) also studied factors influencing maintenance costs at the University of Kaiserslautern in Germany. He concluded that the average effort in staff hours per maintenance task is best explained or predicted by those combined complexity metrics which measure external complexity by information flow and which measure internal complexity by length or number of structural units, i.e., nodes and edges. This means that the complexity of the target software should definitely be considered when estimating the costs of maintenance tasks.

This author has examined these effort estimation equations in the light of reengineering of existing software that could reduce complexity and increase quality, thus lowering maintenance costs. However, there were limits to this.

Complexity could only be reduced by up to 25% and quality could be increased by no more than 33% thus bringing about a maximum effort reduction of 30%! This conclusion is supported by the fact that the product itself is only one of four maintenance cost drivers:

- the product being maintained,
- the environment in which the product is being maintained,
- the maintenance process, and
- the personnel doing the maintenance, (Sneed, 1991).

More recent studies have been made by Lanning and Khoshgoftaar (1994) in relating source code complexity to maintenance difficulty and by Coleman, Ash, Lowther and Oman (1994) to establish metrics for evaluating software maintainability. The latter compute a maintainability coefficient to rate programs as either highly maintainable, i.e., above 0.85, moderately maintainable, i.e., above 0.65, or difficult to maintain, i.e., those under 0.65. Both studies come to the conclusion that complexity and maintainability are related.

In recent years, the first studies have been published on the maintainability of object-oriented software. Chidamber and Kemerer (1994) have developed a set of six metrics for sizing object-oriented software:

- weighted number of methods per class,
- depth of inheritance tree,
- number of subordinate classes per base class,
- coupling between object classes, i.e., the number of messages,
- number of possible responses per message, and
- cohesion between methods in a class, i.e., the number of commonly used data attributes.

These metrics are very important in finding a way to size object-oriented software. Equally important are metrics for determining the complexity of object-oriented software. Wilde and Huitt (1992) have identified those features of object-oriented software, such as polymorphism, inheritance and dynamic binding, which complicate maintenance tasks. They maintain that the unrestricted use of these features tends to drive maintenance costs up by increasing the complexity and decreasing the transparency of the software.

These and the many other studies on the relationship of software product size, complexity, and quality to maintenance costs must be considered when developing a maintenance costing model. This is, however, only the product side. Besides the maintenance process with the steps required to process a change request or error report as described by Mira Kajko-Mattsson (1999), there is also the environment side to be considered, with features of the maintenance environment such as source accessibility, browsing, editing, compiling and testing facilities, communication facilities, influence maintenance costs, etc. Here, studies have been made by

Figure 1: Cost Drivers in Software Maintenance

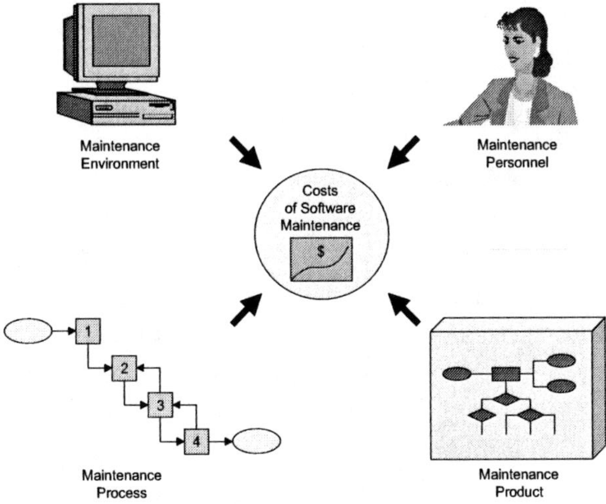

Thadhani (1984), Lambert (1984), and Butterline (1992), which show that maintenance productivity can be increased by up to 37% by improvement to the maintenance environment, in particular the use of a dedicated maintenance workbench as proposed by this author in another paper. (Sneed, 1995).

All of this goes to show how difficult it is to create a comprehensive model for estimating software maintenance costs. It is not only necessary to consider the product to be maintained, but also the environment in which it is to be maintained. Finally, the quality of the maintenance personnel has to be considered. These are the four general cost drivers in software maintenance as depicted in Figure 1.

RESEARCH ON IMPACT ANALYSIS

Most of the previous research on maintenance cost estimation including the work of Putnam (1980), Jones (1991), Boehm (1983) and Nesi (1998) has been devoted to the macro economics of software maintenance, i.e., the prediction of total or annual maintenance costs rather than the micro economics of costing individual maintenance tasks, which requires a completely different approach.

The effort estimation of individual maintenance tasks, whether they be corrections, adaptations, or optimizations, depends on the size of the software impact and the rate of impact. This again is dependent upon an impact analysis. Impact analysis determines which elements of a software system are affected and to what degree they are affected. Thus, maintenance task estimation and impact analysis are highly related. Except for enhancements which do not affect the existing software, impact analysis is a prerequisite of maintenance task estimation.

That relationship was first analyzed by this author in a paper for the International Conference on Software Maintenance in 1995 entitled "Estimating the Costs of Software Maintenance Tasks." Here the author reviews the different methods of sizing software:

- by lines of codes or statements,
- by function-points,
- by data-points, and
- by object-points (Sneed, 1995).

All of these size measurements can be derived either from the existing source code or from the specification documentation by means of automated static analysis. This way there is not only a size of the system as a whole but also a size of each individual component.

The first step in estimating a maintenance task is then to identify which modules, databases, and interfaces are impacted by the planned change. The next step is to subject these artifacts to an automated analysis which returns the size of the impacted artifacts. Then comes the difficult part, that of estimating the degree of change. The change rate is multiplied by the size of the impact domain in lines of code, statements, function-points, data-points, or object-points to obtain the unadjusted size of the maintenance task. This size measure is then adjusted by the complexity factor, the quality factor, and the environment influence factor to derive the adjusted size in whatever is being counted. Once the adjusted size is known, it is possible to convert it by means of maintenance productivity tables or a logarithmic curve into effort in person months.

$$\text{Impact Size} = \Sigma(\text{Artifact Sizes}) * (\text{ChangeRate})$$

$$\text{Adjusted Impact Size} = \text{Impact Size} * \text{ComplexityFactor} * \text{QualityFactor} * \text{InfluenceFactor}$$

$$\text{Effort} = \frac{\text{Adjusted Impact Size}}{\text{Productivity Rate}}$$

Impact analysis is applied to existing software systems to determine which elements of a system are affected by a change to that system. It presupposes that a particular starting point is known. The starting point could be any element of the system—a user interface, a file or database, a module or class, an internal interface – whatever it is that has to be changed. From there, all dependencies are traced to the dependent elements and their dependent elements until the boundaries of the dependency have been reached. The superset of all dependent elements is then referred to as the impact domain. It could possibly be the whole system in the worse case or a small segment thereof in the best case, (Bendifallah & Scacchi, 1987).

The first work on impact analysis was started in the 1980s and documented in a book by Robert Arnold in 1990 with the title *Impact Analysis of Software Systems.* (Arnold, 1990) This early work concentrated on the so-called ripple effect, i.e., by changing one module identifying which modules it calls and which modules call it. The extent of the ripple effect was determined by the branches of the call tree.

Later the ripple effect was extended to data variables to check where they are used and in connection with what other variables. The data was traced backwards from argument to result to argument and forwards from result to argument to result. Later this technique was to prove highly useful in tracking down date fields. Pfleger and Bohner (1990) recognized the significance of impact analysis for software maintenance and extended it to include source code, test cases, design, documents, and requirements. The dependencies between these various software artifacts was depicted in a trace ability graph.

Monroe and Turver (1994) introduced the concept of a ripple propagation graph. Starting with the change request, paths were followed through the documentation libraries to all documents impacted by the change. The result was a probability connection matrix whereby each document was weighted by a probability measure of the degree to which it could be Affected. Thus, not only were the nodes within the impact domain identified, but also their probable degree of impact. The approach taken could possibly be applied to the project discussed here.

Lanubile and Visaggio (1995) continued this work with a study of decision driven maintenance in which they combined documents, programs, and data into a system web. They not only connected documents with one another, but documents with programs and programs with data. They also determined whether the links were conditional or not. Their traceability support system drew upon the case repository, the document libraries, the source libraries, and the database schemas to generate a model of the system which could be analyzed for depending relationships.

Impact analysis research was intensified by the year 2000 data change. Date fields became the objects to which all dependencies had to be traced. The backward trace followed the flow of data from the date field back to the point of input on the system interface, whereas the forward trace followed the data flow from the date field to the point of output in the database or in the user interface. At the 1996 Maintenance Conference, Bohner presented a paper on the role of impact analysis in the software change process, describing an automated impact analysis system for identifying date fields, (Bohner, 1996).

At the same conference, Gallagher (1996) reported on a tool for visualizing impact analysis, graphically depicting the dependencies between C functions and global variables, while McCrickard and Abowd (1996) addressed the impact of change at the architectural level. The interesting aspect of their research was the use of a debugger to dynamically trace dependencies, a theme adapted by the author in source animation.

This other variation of dynamic impact analysis was supplied four years later by the author, who developed an animator for C++ and JAVA classes to trace function calls across class and component boundaries in order to recreate sequence diagrams from the source. These sequence diagrams documented not only the dependencies but also in which sequence they occur, (Sneed, 2000).

Impact analysis is related to at least three other areas of research in software maintenance. These are:

- slicing technique,
- hypertext technique, and
- concept lattices.

An article by Gallagher and Lyle (1991) summarizes the contribution of **slicing** to impact analysis. The authors state,

> We apply program slicing to the maintenance problem by extending the notion of a program slice to a decomposition slice, one that captures all computation on a given variable – data slicing – as well as to one that captures all branches to and from a given procedure – procedural slicing. Using this lattice of single variable or single procedure decomposition slices ordered by set inclusion, we demonstrate how to attain a slice-based decomposition of programs.... Program slicing techniques give an assessment of the impact of proposed modifications...

Another related subject is that of **hypertext**. Hypertext is a technique used to relate distributed information nodes and to navigate through these nodes by means of association. As pointed out by Smith and Weiss in a special issue of the CACM in 1988, "the human mind operates by association. With one item in its grasp, it snaps instantly to the next that is suggested by the association of thoughts, in accordance with some intricate web of trails carried by the cells of the brain." Hypertext documents are linked directly by cross references from one document to another and indirectly by keywords in context which can be associated into the same keywords in other documents, thus resulting in a chain of links through the network. A hypertext document is one in which information is stored in nodes connected by such links (Smith & Weiss, 1988).

Many of the issues involved in hypertext documents are equally applicable to source code. The basic questions are:

- What data model should be used to represent the node network?
- What is the optimal size of a node?
- What types of nodes should there be?
- What nodes should be used as the source of a link?
- What nodes should be used as a destination of a link?
- What types of links should there be?
- Should links have an internal structure?
- How can nodes be aggregated into larger structures?
- How should links be displayed?
- How should link source/destinations be presented on the display?

These and other similar questions have plagued hypertext researchers since 1968 when the first hypertext system was constructed at the Stanford Research Institute (Conklin, 1987).

The third related field is that of **concept lattices**. Concept Analysis is used to identify and group objects with common attributes. A concept is a pair of sets – one being a set of objects – the extent, and the other being a set of attributes – the intent.

It is partitioned into pairs of subsets, each containing a collection of objects sharing common attributes. Each sub concept relation forms a complete partial order over the set of concepts referred to as the concept lattice (Snelting, 1996).

Concept Analysis has been applied to the modularization of programs in design as well as to the identification of potential modules in reengineering (Siff & Reps, 1997). It can also be applied to impact analysis to identify the impact domain. Assuming that attributes of objects to be changed are related to attributes of other objects not being changed, then the subset of objects with attributes related to the changed attributes – the concept lattice – is equivalent to the impact domain. (AnQuetil, 2000).

SOFTCALC MODEL FOR ESTIMATING MAINTENANCE PROJECT COSTS

SoftCalc is a cost estimation tool designed to synthesize and operationalize the research results cited above and to merge them with conventional estimating techniques. It was developed especially for estimating maintenance tasks via a seven step process. (See Figure 2.)

In **Step 1**, an automated audit of the software is conducted to determine its size, complexity, and quality.

In **Step 2**, the impact domain of the software affected by the planned maintenance action is circumscribed.

Figure 2. Maintenance Estimation Process

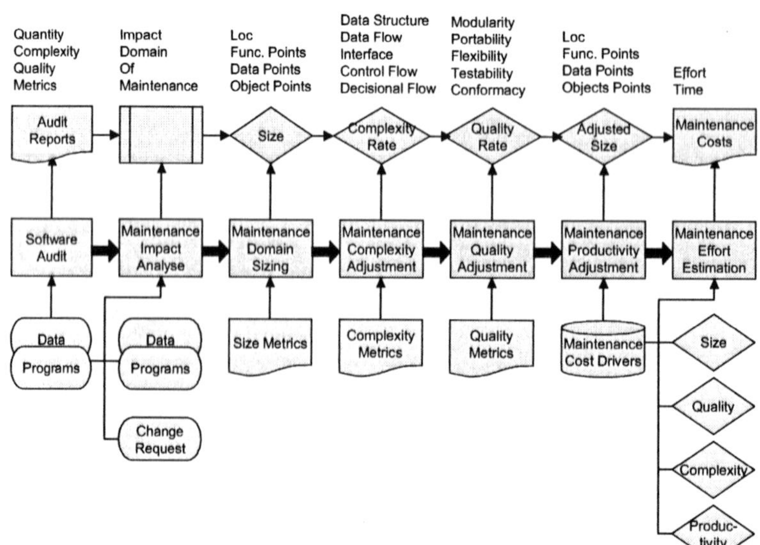

In **Step 3**, the size of the impacted domain is measured in two or more of the following size metrics:

- lines of code,
- statements,
- function-points,
- data-points, and
- object-points.

In **Step 4**, the size measure is adjusted by a complexity factor calculated using a code auditor.

In **Step 5**, the size measure is adjusted by the external and internal quality factors. The latter, which reflects the software maintainability, is also obtained automatically from a code audit.

In **Step 6**, the size measure is adjusted by a productivity influence factor depending on the estimation method.

In **Step 7**, the adjusted size measure is transposed into maintenance effort by means of a productivity table.

Measuring the Size of Software

There are a number of different measurements of software size:

- lines of code,
- statements,
- function-points,
- data-points, and
- object-points,

each of which has advantages and disadvantages depending on the nature of the software.

When dealing with software written in a third generation procedural language such as FORTRAN, COBOL or PL/1, lines of code or statements can be used to measure size. The problem with counting lines is that a statement may be spread out over several lines. The number of lines is often a question of programming style. Therefore, it is better to use statements, especially with a language like COBOL. Each data definition and each verb denotes a statement. This way ON and END lines are not counted. The same applies to PL/1, where each statement is terminated by a semicolon. Only in FORTRAN is there really a relatively fixed relation between statements and lines of code. The only thing which speaks for lines of code is that they are easy to count. However, with a simple tool, it is just as easy to count statements. In any case, when dealing with existing software, it is recommended to count statements.

In the case of software systems written in a 4th generation language such as FOCUS, ADS-ONLINE, ABAP or NATURAL, it is best to measure the size of the existing system in data-points. Data-points are a measure for the number of data elements processed, the number of user views, the number of data fields in the views, and the complexity of the interfaces (Sneed, 1990). Data elements and data fields are weighted with 1, relational tables with 4, database accesses with 2, and user views with 4. User views are adjusted by a complexity factor of 0,75 to 1.25. The sum of the adjusted data-points is a reliable indicator for the size of 4GL systems which connect user interfaces to relational databases.

For object software, the most appropriate measure of size is the object-point. Object-points are derived from the class structures, the messages, and the processes or use cases (Sneed, 1995).

Class-Points	=	(number of class attributes ´ 1)
	+	(number of class relationships ´ 2)
	+	(number of class methods ´ 3)

for each class.

Message-Points	=	[(number of parameters ´ 1)
	+	(number of senders and receivers ´ 2)]
	´	complexity rate for each message.

Process-Points	=	(process multiplication factor of 1 to 6 depending on the process type)
	+	(number of process variations ´ 2)

In the case of classes, the number of class-points for each class is adjusted by the reusability rate. It is important to note that inherited attributes and methods are only counted once, at the base or super class level.

Class-points and message-points can be derived automatically from the code itself. Process-points must be counted in the user documentation. The object-point count is the sum of the class-points, message-points, and process-points. As such, it is a union of internal and external views of the software.

Function-points are a universal measurement of software size for all kinds of systems. They are obtained by counting system inputs, system outputs, system queries, databases, and import/export files (Albrecht & Gaffney, 1983). System inputs are weighted from 3 to 6; system outputs are weighted from 4 to 7; system queries are weighted from 3 to 6; databases are weighted form 7 to 15, and import/export files are weighted from 7 to 10. It is difficult to obtain the Function-point count by analyzing the code. The question is what is an input and an output. Counting READ's and WRITE's, SEND's and RECEIVE's, DISPLAY's and ACCEPT's is misleading, since there is no direct relationship between program I/O operations and system inputs and outputs. The actual number of system inputs and outputs is a subset of the program inputs and outputs. This subset has to be selected manually from the superset derived automatically by comparison with the system documentation. The database function-point count can be derived by counting physical databases or relational tables in the database schema. Import/Export Files can be found in the Interface Documentation. In light of the strong reliance on documentation rather than code, there is much more human effort involved in counting function-points, unless this count can be obtained from a CASE Repository.

Independently of what unit of measure is used, the result of the system sizing task is an absolute metric representative of the size of the system as a whole. The next step is to determine what portion of this size measurement is affected by the planned maintenance operation. This is the goal in impact analysis.

Determining the Size of the Impact Domain

Once the size of the system under maintenance has been quantified, it is only necessary to update the size figures after each new release. Maintenance action requests come in the form of error reports and change requests as well as in the form of reengineering project proposals. These requests can be placed in the traditional categories of:

- corrective,
- adaptive,
- perfective, and
- preventive maintenance (Lientz & Swanson, 1981).

In order to estimate the costs of a maintenance action, it is first necessary to determine the size of the impact domain of that action, i.e., the proportion of the entire software affected. This study is referred to as impact analysis and has been covered in the literature by Robert Arnold (1989), Malcom Munroe (1994) and

Figure 3. Sizing the Impact Domain

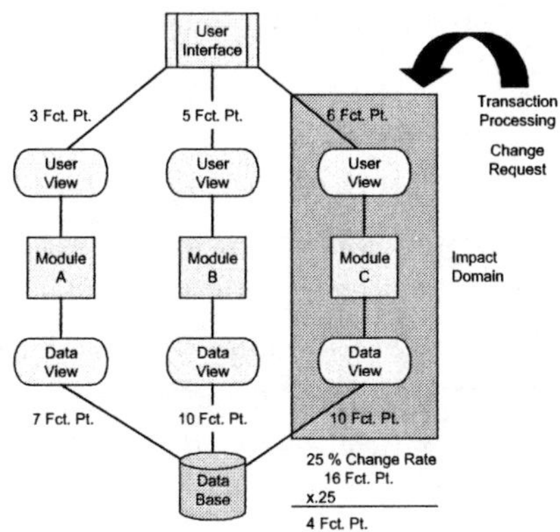

others. The object of the impact analysis depends on how the software has been sized. If it is sized in function-points, it will be necessary to analyze the system documentation with the following questions in mind:

- What inputs and outputs must be added?
- What inputs and outputs must be altered?
- What queries must be added?
- What queries must be altered?
- What import/export files must be added?
- What import/export files must be altered?
- What files or databases must be added?
- What files or databases must be altered?

If inputs, outputs, queries, import/export files, or databases are to be added, then their function-points must be counted in full. If they are to be altered, then the question is to what extent. Here, the analyst must exercise judgement in coming up with a percent change factor. One method is to count the number of data elements to be changed or added relative to the sum of the data elements in that particular input/output interface or file. (See Figure 3.)

If the system is sized in object-points, it is the goal of the analyst to first establish which classes have to be added or altered. If a class is to be added, then its attributes, relationships, and methods are counted in full. If it is to be altered, then only those attributes, relationships and methods which are affected by the change are counted. The same applies to messages. New messages are counted in full; altered messages are counted relative to the proportion of parameters, senders, and receivers affected. Finally, new processes are counted in full, but altered processes are only counted if there is a new variant to be implemented. The sum of the impacted object-points is then some portion of the total number of object-points. This will be the most likely method of sizing maintenance actions on object-oriented software.

In case the system has been sized in data-points, the focus of the impact analysis is on the data model. Here, it is required to answer the following questions:

- What data entities must be added?
- What data elements in existing data entities must be added or changed?
- What relationships between existing data entities must be added or changed?
- What user views must be added?
- What data fields in existing user views must be added or changed?
- What relationships between existing data entities and existing user views must be added or altered?

New entities and user views are counted in full. Altered entities and user views are counted in proportion to the number of their attributes and relationships affected by the planned maintenance action. The result is some portion of the total system size in data-points.

Finally, existing systems can always be sized in terms of lines of code or statements. The object of the impact analysis is, in this case, the code. The problem is to determine first what components are affected by the change and then to determine what percentage of the statements, i.e. data declarations and procedural instructions, are affected. Of all the sizing approaches used in impact analysis this is surprisingly the most difficult, because of the fine granularity. Either one uses some kind of sophisticated slicing technique to identify precisely which program paths and which data must be altered, or one has to make an educated guess. The former method is accurate but costly, the latter cheap but inaccurate. Thus, counting statements affected by maintenance is no easy task. It may be better to size maintenance impact domains in Function-Points, Object-Points or Data-Points and then to transpose this count into a statement count based on a size relationship factor as that published by Capers Jones (1991).

Adjusting the Size of the Maintenance Impact Domain by the Complexity Factor

From the voluminous amount of literature on the subject it is obvious that the complexity of the software has a definite effect on the maintenance effort. The question is what complexity metrics to use. In his book on software complexity, Horst Zuse (1990) describes no less than 139 different complexity metrics. Therefore, anyone faced with the task of assessing complexity must somehow make a selection.

In the development of its audit tools for Assembler, PL/1, COBOL, and NATURAL, the author's company has reduced the many possible metrics to five basic ones which cover the main dimensions of software:

- data complexity,
- data usage complexity,
- control flow complexity,
- decisional complexity, and
- interface complexity.

Data complexity is derived using Chapin's Q-complexity measure (Chapin, 1979). Data usage complexity is based on Elshof's referential frequency complexity (Elshof, 1976). Control flow complexity is McCabe's cyclomatic complexity. (McCabe, 1976). Decisional complexity is an adaptation of McClure's decisional nesting metric (McClure, 1981). Interface complexity is based on Henry and Kafura's system cohesion metric (Henry & Kafura, 1981). All of the metrics are converted onto a rational scale of 0 to 1 with 0.5 being the median complexity, 1 the highest acceptable, and 0 the lowest possible. The resulting coefficient is then added to 0.5, the median complexity to give a multiplication factor with a base of

1. This factor is used to adjust the size of the impact software by complexity. Thus, a complexity factor of 1.1 will cause a 10% increase in the size of the impact domain.

Complexity-Adjusted-Size =
Raw-Size * Complexity-Factor

Adjusting the Size of the Maintenance Impact Domain by the Quality Factor

The quality of software has a definite influence on productivity, not only in development, where it is an objective to be achieved, but also in maintenance, where it is not only an objective to be achieved, but also a constraint to be overcome. In development, quality is definitely a cost driver. The higher the quality goals, the higher the costs. In maintenance, this is not so simple. On the one hand, quality in terms of reliability, security, performance, usability, etc., must be maintained at same level if not enhanced, otherwise maintenance degradation sets in. On the other hand, quality in terms of portability, flexibility, readability, modularity, interoperability, etc., is a constraint. If the quality of the existing software is low, the costs of implementing maintenance actions will be higher. If it is high, then these costs will be lower (Boehm, 1984).

For this reason, the quality factor has to be computed twice. First, the external quality characteristics – reliability, security, integrity, usability, time efficiency, and space efficiency, are rated on a scale from 0.5 to 1.5, whereby 1 is the industry median. The highest possible quality is 1.5, the lowest acceptable is 0.5. The system external quality factor is the average of the individual external quality factors.

In a second step, the internal quality characteristics – portability, flexibility, readability, modularity, data independence, and interoperability are rated on a similar scale to the program complexity with a lower bound of 0, an upper bound of 1 and a median of 0.5. The average of these tool-derived internal metrics is then subtracted form 1, i.e., inverted, and added to 0.5 to give a multiplication factor as expressed in the rule:

MF = (1 – Average-Quality) + 0.5

In this way, an average quality above 0.5 will reduce the size of the impact area, whereas a below average quality will increase the size. The problem here is, of course, in defining what is an average grade for maintainability, i.e., what is an average cyclomatic complexity, an average nesting level or an average number of interfaces?

The size measure in statements, data-points, object-points, or function-points is adjusted by multiplying it first by the external quality factor and then by the internal quality factor, both being independent of one another. For example, if the function-point count of the software impacted is 20, the external quality factor 1.1, and the inverted internal quality factor 0.8, the adjusted function-point count will be:

20 * 1.1 * 0.8 = 17

Adjusting the Size of the Maintenance Costs by the Project Influence Factor

The project influence factor is computed differently by each estimation method. In the COCOMO method it is the product of 16 project cost drivers on a scale of 0.4 to 2.0. The function-point method weighs 14 factors on a scale of 0 to 5 and adds the sum multiplied by 0.01 to 0.65 (Low & Jeffery, 1990). The data-point method weighs 10 factors on a scale of 1 to 5 and subtracts the sum multiplied by 0.01 from 1.25. The object-point method assigns weights from 0 to 4 to 10 different factors and subtracts the sum multiplied by 0.01 from 1, meaning that the number of object-points can only be adjusted downwards.

Multiplying the system size by the project influence factor renders the final adjusted size of the impact domain.

Adjusted-Size = Raw-Size * Complexity-Factor * Quality-Factor * Influence-Factor.

Computing the Maintenance Effort

The relationship between the adjusted size of the impact domain and the effort required to perform the maintenance action is derived from a maintenance productivity table. In general, it can be said that maintenance productivity is about half of development productivity. If the development productivity is 50 statements per day, then the maintenance productivity will be 25. Likewise, a function-point productivity of 10 function-points per developer man-month means 5 function-points per maintenance man-month.

One reason for this is the additional effort required to comprehend the software. A second reason is the additional effort needed to integrate and retest the altered components. Together, these factors drive the effort up by some 100%. This conclusion has been made by several independent studies of maintenance productivity (Grady, 1987).

Essentially, it is up to each software maintenance shop to calibrate its own maintenance productivity curves reflected in the tables based on its own experience. This will make the estimating process more accurate.

CASE STUDY OF APPLYING THE SOFTCALC MODEL

In the early 1990s, the author's company had just completed the reengineering of nine Basic Assembler programs with 7,500 statements in a financial application system to COBOL-85. The new COBOL versions were due to go into production by the end of the year. However, due to changes in the tax laws, the programs had to be altered before going into production. The changes were first built by the customer programmers into the original Assembler programs and tested. Fortunately, they were also marked, so it was possible to identify them.

After the original Assembler programs had been validated, the statements in the COBOL programs which corresponded to the marked Assembler statements were identified. The changes were significant so that in all some 372 Assembler statements or 5% of the total code had been altered. This corresponded to precisely 410 COBOL statements.

The complexity measurements of the Assembler code derived by the ASMAUDIT tool were:

Data Complexity	= 0.743
Data Usage Complexity	= 0.892
Control Flow Complexity	= 0.617
Decisional Complexity	= 0.509
Interface Complexity	= 0.431

giving an average complexity for the Assembler programs of 0.636. This was transposed to a complexity factor of 1.14. (See Figure 4.)

Figure 4. Assembler System Metrics

PROGRAM LANGUAGE: 370-ASSEMBLER	DATE: 12.10.93
SYSTEM NAME: COUPONS	PAGE: 001
QUANTITY METRICS	
PROCEDURE SIZE METRICS	
NUMBER OF FILES/DATABASE	11
NUMBER OF DATA OBJECTS	18
NUMBER OF DATA DECLARED	2473
NUMBER OF DATA REFERENCES	1805
NUMBER OF ARGUMENTS	3008
NUMBER OF RESULTS	2632
NUMBER OF PREDICATES	1428
NUMBER OF COPY/MACROS	38
NUMBER OF DATA-POINTS	712
PROCEDURE SIZE METRICS	
NUMBER OF STATEMENTS	7521
NUMBER OF MODULES	9
NUMBER OF BRANCHES	2933
NUMBER OF GO TO BRANCHES	1254
NUMBER OF SUBROUTINE CALLS	207
NUMBER OF MODUL CALLS	19
NUMBER OF DATA REFERENCES	4961
NUMBER OF FUNCTION POINTS	55
NUMBER OF LINES OF CODE	9994
NUMBER OF COMMENT LINES	1093
COMPLEXITY METRICS	
DATA COMPLEXITY	0,743
DATA FLOW COMPLEXITY	0,892
INTERFACE COMPLEXITY	0,421
CONTROL FLOW COMPLEXITY	0,617
DECISIONAL COMPLEXITY	0,509
INTERNAL COMPLEXITY	0,636
QUALITY METRICS	
MODULARITY	0,339
PORTABILITY	0,214
MAINTAINABILITY	0,364
TESTABILITY	0,472
CONFORMITY	0,702
INTERNAL QUALITY	0,418

System Summary Report

The complexity measurements of the COBOL code derived by the COBAUDIT tool were:

Data Complexity	= 0.744
Data Usage Complexity	= 0.811
Control Flow Complexity	= 0.529
Decisional Complexity	= 0.506
Interface Complexity	= 0.421

giving an average complexity for the COBOL programs of 0.602. This was transposed to a complexity factor of 1.10. (See Figure 5.)

Adjusting the size of the Assembler Impact Domain gave a statement count of 424. Adjusting the size of the COBOL Impact Domain gave a statement count of 451, which was still greater than that of the Assembler due to the additional statements created by the automatic conversion.

The external quality of the code remained constant at 0.05. However, the internal quality had risen as a result of the reengineering from a previous 0.418 for the Assembler to 0.473 for the COBOL version. When inverted, this gave the multiplication factor of 1.08 for Assembler and 1.03 for COBOL. This resulted in a quality-adjusted COBOL statement count of 415, as opposed to the quality

Figure 5. COBOL System Metrics

PROGRAM LANGUAGE:	370-ASSEMBLER	DATE:	12.10.93
SYSTEM NAME:	COUPONS	PAGE:	001

QUANTITY METRICS	
PROCEDURE SIZE METRICS	
NUMBER OF FILES/DATABASE	11
NUMBER OF DATA OBJECTS	18
NUMBER OF DATA DECLARED	2473
NUMBER OF DATA REFERENCES	1805
NUMBER OF ARGUMENTS	3008
NUMBER OF RESULTS	2632
NUMBER OF PREDICATES	1428
NUMBER OF COPY/MACROS	38
NUMBER OF DATA-POINTS	712
PROCEDURE SIZE METRICS	
NUMBER OF STATEMENTS	7521
NUMBER OF MODULES	9
NUMBER OF BRANCHES	2933
NUMBER OF GO TO BRANCHES	1254
NUMBER OF SUBROUTINE CALLS	207
NUMBER OF MODULE CALLS	19
NUMBER OF DATA REFERENCES	4961
NUMBER OF FUNCTION POINTS	55
NUMBER OF LINES OF CODE	9994
NUMBER OF COMMENT LINES	1093
COMPLEXITY METRICS	
DATA COMPLEXITY	0,743
DATA FLOW COMPLEXITY	0,892
INTERFACE COMPLEXITY	0,421
CONTROL FLOW COMPLEXITY	0,617
DECISIONAL COMPLEXITY	0,509
QUALITY METRICS	
MODULARITY	0,339
PORTABILITY	0,214
MAINTAINABILITY	0,364
TESTABILITY	0,472
CONFORMITY	0,702
INTERNAL QUALITY	0,418

adjusted Assembler statement count of 458. At this point, the size of the COBOL impact domain was still larger.

The project influence factors had also improved as a result of the reengineering. It was now possible to edit, compile, and test the programs on a PC-workstation. The product of the cost drivers was 1.26 for the Assembler Code. For the COBOL Code, it was 0.98. Thus, whereas the final adjusted Assembler statement count came out to be 577, the adjusted COBOL statement count was 21% less at 456.

At a productivity rate of 20 Assembler statements or 30 adjusted statements per man-day, it took some 19 man-days to adapt and retest the original Assembler programs. On the COBOL side, it took only 10 man-days to duplicate the changes and retest the programs giving a maintenance productivity rate of 45 adjusted statements or 41 real statements per man-day. The effort of specifying the changes was not included on either side, but it would have been the same for both.

This is a good example of how reengineering can reduce maintenance costs. Even though the automatic conversion of Assembler to COBOL resulted in more code, it was still possible to alter the COBOL code quicker and cheaper than the original Assembler code. This was due primarily to the better support offered by the Microfocus COBOL Workbench which underlies the impact of the environment on maintenance productivity.

MAINCOST – A TOOL FOR ESTIMATING INDIVIDUAL CHANGES

The tool MAINCOST has been developed by the author in the past year to estimate maintenance tasks for a large Vienna software house developing and maintaining stock brokerage application for banks. The tool combines impact analysis with maintenance cost estimation. As such, it consists of two components – CodeScan and CodeCalc. CodeScan performs the impact analysis on the source code repository whereas CodeCalc computes the effort based on the source itself. In dealing with large, complex systems, it is not possible to analyze impact directly within the source. The key entities and their relationships must first be extracted from the source code and stored in a database. It could be a relational or a graphical database. The important factor is that all tuples of the type base entity : relation : target entity are storable.

MAINCOST uses the **SoftRepo** Repository for representing the structure and contents of object-oriented application systems implemented in C++ and Java. In **SoftRepo** there are altogether some 16 relational tables stored in UDB2. These tables depict the following relationships:

Component <belongs to> System
Component <contains> Modules

Module<contains>Modules (Includes)
Module<contains>Classes
Module<contains>Attributes
Module<contains>Functions
Module<processes>DB-Tables
Module<processes>Panels
Class<inherits from>Class
Class<contains>Classes
Class<contains>Attributes
Class<contains>Functions
Function<invokes>Functions
Function<receives>Parameters
Function<returns>Values
Function<accesses>DB-Tables
DB-Table<contains>Attributes (See Figure 6.)

When entering MAINCOST, the user is presented with a list of software entity types to choose from. The change request may apply to a class, a function, a parameter, an attribute, a DB-table, or a panel. It is up to the user to identify which

Figure 6: SoftRepo Repository Model

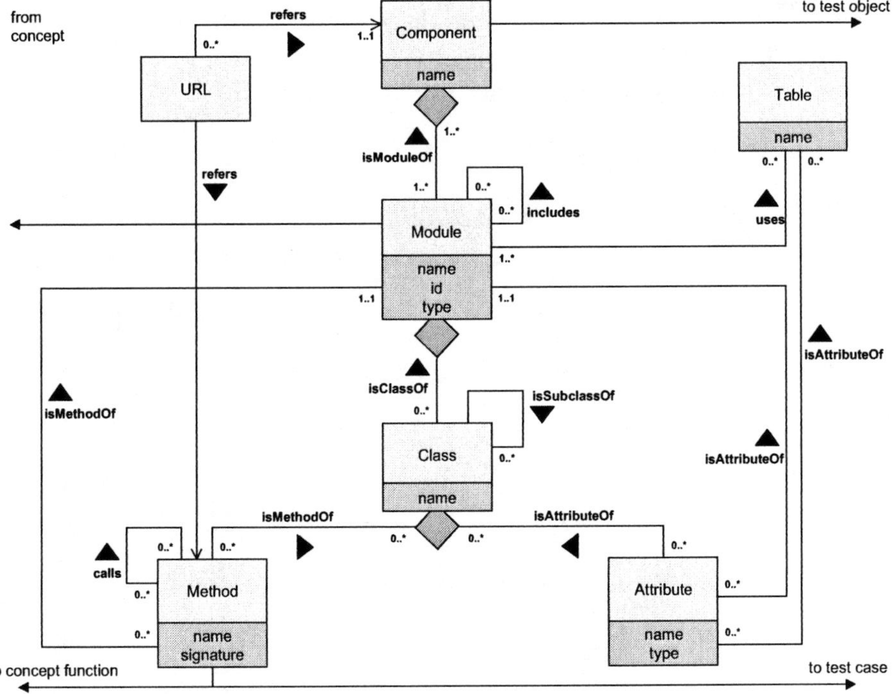

type of software entity is affected. The tool comes back with a list of all entity occurrences of that type, e.g., a list of all classes. Then the user selects that entity or those entities directly affected by the change. These are the first level impacted source entities. They are passed as parameters to a backend component which then searches through the repository for all entities to which they are related. These are the second level impacted entities. This recursive search algorithm goes on up to a maximum closure level defined by the user.

The result of the impact analysis is a tree of impacted source elements identified by name and level. The level corresponds to their distance from the base element, measured in hierarchical levels. This tree is the output of CodeScan and the input to CodeCalc. CodeCalc processes the tree node for node, searching out the source text of the particular entity. Each of the source texts referenced is measured in terms of its size, quality, and complexity. For the size statements, data-points, function-points, and object-points are counted. Complexity is measured here in terms of data structure, data flow, data access, interface (fan-in/fan-out), control flow (cyclomatic complexity), decisional depth, branching level, inherence depth, and language volume. Quality is measured in terms of portability, modularity, flexibility, testability, readability, maintainability, security, and conformity to standards. These metrics for each impacted element are stored in a table and accumulated (see Figure 7).

The sizes of the impacted elements are adjusted by two factors. One is the tree level or distance from the point of impact. The size metrics are divided by the level number thus reducing their impact depending on their distance from the base element. The other factor is the change rate given by the user. It is a percentage of the total size adjusted by the distance to the point of impact. It takes only that percentage of the size affected by the change.

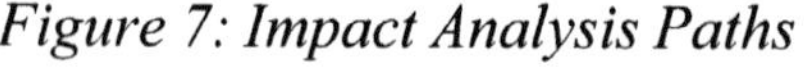
Figure 7: Impact Analysis Paths

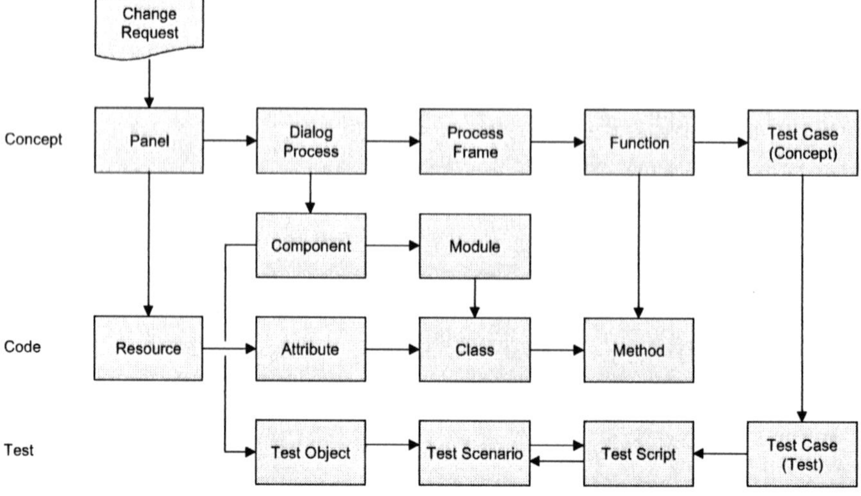

The resulting adjusted size is then adjusted again by the complexity and the quality of that particular source element. In the end, the adjusted sizes are aggregated to give a total adjusted size of the impact domain. It is this total size which is then compared to the existing productivity tables to be converted via interpolation into a number of person-months based on the experience with previous maintenance projects. It is important that the maintenance productivity table be kept up to date. So after every completed maintenance task, the size of the actual impact domain and the actual effort flow back into the productivity table by means of a change analysis and the effort reporting system (see Figure 8).

MAINCOST has proven to be a very effective tool in terms of time and cost estimates. A routineered project planer can start an estimation and receive a range of estimated efforts within minutes. That means, a customer requiring an offer to implement some change or enhancement can get an answer within a few hours. This is only possible with the support of an extensive software repository, impact analysis, tools, source measurement tools and, last but not least, a set of accurate productivity statistics.

FUTURE DIRECTIONS IN MAINTENANCE COST ESTIMATION

Software maintenance encompasses a wide variety of discrete tasks which individually cause little costs, but which, in their sum, account for a significant effort Chapin and Hale, 2001. As long as users were developing and maintaining their own

Figure 8. Maintenance Task Estimation

Requester:	Bank of Austria	System:	GEOS
Change Request:	Alter_Order_Map	Date:	2001.01.22

Level	Type	Artifact	Lines	Stmts	Fct. Pts	Obj. Pts.	Test Pts.	Comp	Qual	Impact Rate
Concept	Panel	Order_Map	39	39	7	21	-	0.420	0.602	10%
Concept	Dialog	Order_Entry	67	67	13	39	-	0.491	0.598	10%
Concept	Frame	Order_Process	208	208	20	112	-	0.523	0.570	5%
Concept	Funct	Status_check	92	92	6	28	-	0.318	0.635	20%
Concept	TestCase	TF_OE_011	9	9	-	-	4	-	-	50%
Concept	TestCase	TF_OE_013	7	7	-	-	3	-	-	50%
Code	Comp	Procord	2420	1680	51	803	-	0.561	0.501	2%
Code	Module	Procord1	860	560	19	240	-	0.573	0.499	5%
Code	Class	Order	97	42	-	28	-	0.519	0.538	10%
Code	Method	Check_Status	36	21	-	-	-	-	-	40%
Test	TestCase	TF_OE_011	12	8	-	-	4	-	-	50%
Test	TestCase	TF_OE_013	10	6	-	-	3	-	-	50%
Test	TestObj	Procord	219	150	-	-	72	-	-	2%
Test	Scenario	Test_Orders	66	45	-	-	21	0.592	0.542	10%
Test	Script	Test_Status	212	158	-	-	69	0.571	0.531	20%
		15	4354	3092	116	1271	176	0.507	0.557	3%

software applications, there was no pressing need to accurately estimate the maintenance costs. The budget was adjusted to cover whatever costs occurred. This is one of the main reasons why IT costs have gone out of control and why user managers have turned to outsourcing or purchasing standard software as an alternative. From outsourcing, they expect to have better control of the costs. From standard products, they hope to push the responsibility for maintenance off on the supplier. All of the indicators show that maintenance and operations costs now eat up more than 60 % of the IT budget (Pigorski, 1997).

In the case of outsourcing, it is the responsibility of the outsourcing contractor to control the costs of maintenance. In the case of standard software products, e.g., enterprise resource planning systems, or off the shelf components, the supplier of the product has the burden of estimating maintenance costs and the user has the option to reject the offer. In both cases, operational costs are more transparent. It is not so much that the maintenance work is cheaper, but the fact that maintenance costs can be kept under control. That in itself is a great advantage to the enterprise management.

Both alternatives require an accurate cost estimation process. In the case of outsourcing, the outsourcer is interested in keeping his customer satisfied while at the same time covering his costs. In the case of standard software, the supplier is interested in getting the customer to pay for all adaptations and enhancements to his product. The only way to achieve these objectives is through accurate cost estimation.

Therefore, the demand for more systematic, metric-based estimation methods will continue to grow. There is a definite need for better data upon which to base the estimates, as well as for better tools to support the estimates. One can expect to have significant progress in the near future, but the key issue remains the statistical data base. More empirically validated data is needed to relate the volume of maintenance tasks to human effort. This is at the moment the greatest obstacle to using automated estimation tools such as the one described here.

REFERENCES

Albrecht, A. J. & Gaffney, J. E. (1983). Software function, source lines of code, and development effort prediction – A software science validation. *IEEE Transactions on S. E.,* Vol. SE-9, No. 6, p. 639.

AnQuetil, N. (2000). A comparison of graphs of concept for reverse engineering, *In Proceedings of 8th IWPC-2000,* Limerick, pp. 231-240. New York: *IEEE Press.*

ANSI/IEEE. (1998). Standard 1219 – Standard for software maintenance, *IEEE Standards,* Vol. 2. New York: IEEE Press.

Arnold, R. (1989). Software impact analysis, Tutorial at *IEEE Conference on Software Maintenance,* Orlando, Fl.

Arnold, R. (1990). Impact analysis of software systems, New York: McGraw-Hill, p. 11-29.

Bendifallah, S. & Scacchi, W. (1987). Understanding software maintenance work, *IEEE Trans. on S.E.,* Vol. 13, No. 3, p. 311-334.

Boehm, B. (1981). Software engineering economics, Englewood Cliffs, N.J: Prentice-Hall.

Boehm, B. (1983). Economics of software maintenance. In Proceedings of the Software Maintenance Workshop, Monterey, CA: *IEEE Press,* p. 9.

Boehm, B. (1984) Software engineering economics. *IEEE Trans. on S. E.,* Vol. SE-10, No. 1, p. 4.

Boehm, B. (1988). Understanding and Controlling Software Costs. In *IEEE Trans. on S. E.,* Vol. 14, No. 10.

Bohner, S. (1996). Impact analysis in the software change process – A year 2000 Perspective, In *Proceedings of ICSM,* Monterey, p. 42-51. CA: IEEE Press.

Butterline, M. A. (1992). From mainframe to workstations – Offloading application development, p. 59. Boston, MA: QED Publishing Group.

Card, D., Cotnoir, D. & Goorevich, C. (1987). Managing software maintenance cost and quality. In Proceedings of Conference on Software Maintenance, Austin TX, p. 145. Monterey, CA: IEEE Press.

Chapin, N. (1979). A measure of software complexity. In *Proceedings of Natural Computer Conference,* Chicago.

Chapin, N. & Hale, J.A.O. (2001). Types of software evolution and software maintenance. *Journal of Software Maintenance,* Vol. 13, No. 1, p. 3-29.

Coleman, D., Ash, D., Lowther, B. & Oman, P. (1994). Using metrics to evaluate software system maintainability, *IEEE Computer,* p. 44.

Conklin, E.J. (1987). Hypertext – An introduction and survey, *IEEE Computer,* Vol. 2, No. 9, p. 17-41.

Elshof, J. (1976). An analysis of commercial PL/1 Programs, *IEEE Trans. on S. E.,* Vol. SE-2, No. 3.

Gallagher, K. (1996). Visual Impact Analysis, In *Proceedings of ICSM,* p. 52-58. Monterey, CA: IEEE-Press.

Gallagher, K.B. & Lyle, J.R. (1991). Using program slicing in software maintenance, *IEEE Trans. on S.E.,* Vol. 17, No. 8, p. 751-761.

Grady, R. (1987). Measuring and managing software maintenance. In *IEEE Software,* Vol. 4, No. 9, p. 35 to 45

Henry, S. & Kafura, D. (1981). Software structure metrics based on information flow, *IEEE Trans. on S. E.,* Vol. SE-7.

Jones, C. (1991). *Applied Software Measurement,* McGraw-Hill, New York.

Kajko-Mattsson, M. (1999). Maintenance at ABB: The change execution process. In *Proceedings of IEEE-ICSM 1999*, Oxford, UK, *IEEE Computer Society Press,* p. 307.

Kemerer, C. & Chidamber, S. (1994). A Metrics Suite for Object-Oriented Design, *IEEE Trans. on S. E.,* Vol. 20, No. 6, p. 476.

Lambert, G. (1984). A comparative study of system response time on programmer developer productivity, *IBM Systems Journal,* Vol. 23, No. 1, p. 36.

Lanning, D. & Khoshgoftaar, T. (1994). Modeling the relationship between source code complexity and maintenance difficulty. In *IEEE Computer,* p. 35.

Lanubile, F. & Visaggio, G. (1999). Decision-driven Maintenance, *Journal of Software Maintenance,* Vol. 7, No. 7, p. 91-116.

Lientz, B. & Swanson, E. B. (1981). Problems in application software maintenance, *Communications of the ACM,* Vol. 24, No. 11, p. 763.

Low, G. & Jeffery, D. R. (1990). Function-points in the estimation and evaluation of the software process, *IEEE Trans. on S. E.,* Vol. 16, No. 1, p. 64.

McCabe, T. (1976). A complexity metric. In *IEEE Trans. on S. E.,* Vol. 2, No. 4, p. 308.

McClure, C. (1981). *Managing Software Development and Maintenance,* New York: Van Nostrand Reinhold.

McCrickard, R. & Abowd, G. (1996). Assessing the impact of changes at the architectural level – A case study on graphical debuggers. In Proceedings of ICSM, Monterey, CA: IEEE Press, , p. 59-68.

Munroe, M. & Turver, R. (1994). Early impact analysis technique for software maintenance, *Journal of Software Maintenance,* Vol. 6, No. 1, p. 35.

Nesi, P & Querci, T. (1998). Effort estimation and prediction of object-oriented systems, *Journal of Systems and Software,* Vol. 42, No. 2, p. 89-112.

Pfleger, D. & Bohner, S. (1998). A framework for software maintenance metrics, In *Proceedings of ICSM-1990*, San Diego, p. 320-327. Monterey, CA: IEEE Press.

Pigorski, T. (1997). *Practical Software Maintenance,* p. 29. New York: John Wiley & Sons.

Putnam, L. (1980). Software cost estimating and life cycle control. *IEEE Tutorial,* Los Almitos, p. 24. Monterey, CA: IEEE Computer Society Press.

Rombach, D. (1987). A controlled experiment on the impact of software structure on maintainability, *IEEE Trans. on S. E.,* Vol. SE-13, No. 3, p. 344.

Siff, M. & Reps, T. (1997). Identifying modules via concept analysis. In *Proceedings of ICSM-1997,* Bari, p. 170-179. Monterey, CA: IEEE Press.

Smith, J.B. & Weiss, S.F. (1988). Hypertext. In *Communications of the ACM,* Vol. 31, No. 7, p. 816-819.

Sneed, H. (1990). Die Data-Point-Methode. In *ONLINE, ZfD,* No. 5/90, p. 48.

Sneed, H. (1995). Estimating the costs of object-oriented software, In *Proceedings of Software Cost Estimation Seminar,* Systems Engineering Ltd., Durham, U.K.

Sneed, H. (1995). Estimating the costs of software maintenance tasks. In Proceedings of ICSM-1995, Opio, France, p. 168-181. IEEE Press.

Sneed, H. (1995). Implementation of a software maintenance workbench at the Union Bank of Switzerland. In *Proceedings of Oracle Conference on System Downsizing,* Oracle Institute, Munich, Germany.

Sneed, H. (2000). Source animation of C++ programs, In *Proceedings of 8th IWPC,* IEEE Press, Limerick, p. 179-190S.

Snelting, G. (1996). Reengineering of configurations based on mathematical concept analysis, *ACM Trans. on S.E. and Methodology,* Vol. 5, No. 2, p. 146-189.

Thadhani, A. (1984). Factors affecting programmer productivity during application development." *IBM Systems Journal,* Vol. 23, No. 1, p. 19.

Turver, R. & Munroe, M. (1994). An early impact analysis technique for software maintenance. *Journal of Software Maintenance,* Vol. 6, No. 1, p. 35-52.

Vessey, I. & Weber, R. (1983). Some factors affecting program repair maintenance: An empirical study. In *Communications of the ACM,* Vol. 26, No. 2, p. 128.

Wilde, N. & Huitt, R. (1992). Maintenance support for object-oriented programs, *IEEE Trans. on S. E.,* Vol. 18, No. 12, p. 1038.

Zuse, H. (1990). *Software Complexity – Measure and Methods,* Berlin: De Gruyter Verlag.

Chapter IX

A Methodology for Software Maintenance

Macario Polo, Mario Piattini, Francisco Ruiz
Escuela Superior de Informatica
Universidad de Castilla - La Mancha, Spain

Software maintenance is the most expensive stage of the software life cycle. However, most software organizations do not use any methodology for maintenance, although they do use it for new developments. In this article, a methodology for managing the software maintenance process is presented.

The methodology defines clearly and rigorously all the activities and tasks to be executed during the process and provides different sets of activities for five different types of maintenance (urgent and non-urgent corrective, perfective, preventive, and adaptive). In order to help in the execution of tasks, some techniques have been defined in the methodology. Also, several activities and tasks for establishing and ending outsourcing relationships are proposed, as well as several metrics to assess the maintainability of databases and their influence on the rest of the Information System.

This methodology is being applied by Atos ODS, a multinational organization among whose primary business activities is the outsourcing of software maintenance.

INTRODUCTION

Software Maintenance has been traditionally the most expensive stage of the software Life Cycle (see Table 1), and it will continue to grow and become the main work of the software industry (Jones, 1994). In fact, new products and technologies need to increase maintenance efforts in corrective and perfective (for hypertext maintenance, for example, as reported in Brereton, Budgen, & Hamilton, 1999), as in adaptive (for adapting old applications to new environments, as client/server as discussed in Jahnke and Wadsack, 1999). With this in mind, it is natural that software evolution laws announced by Lehman (1980) have been recently confirmed (Lehman, Perry, & Romil, 1998).

Table 1. Evolution of maintenance costs.

Reference	Date	% Maintenance
Pressman (1993)	1970s	35%-40%
Lientz and Swanson (1980)	1976	60%
Pigoski (1997)	1980-1984	55%
Pressman (1993)	1980s	60%
Schach (1990)	1987	67%
Pigoski (1997)	1985-1989	75%
Frazer (1992)	1990	80%
Pigoski (1997)	1990s	90%

In spite of this, one study conducted by Atos ODS in Europe has shown that most software organizations do not use any methodology for software maintenance, although they do use it for new developments. This usage is really surprising, above all if we take into account that 61% of the professional life of programmers is devoted to maintenance work, and only 39% to new developments (Singer, 1998). So, we agree with Basili et al. (1996) the statement sense that "we need to define and validate methodologies that take into account the specific characteristics of a software maintenance organization and its processes." Furthermore, as these same authors express, the improvement of the maintenance process is very interesting due to the large number of legacy systems currently used.

Usual solutions for software maintenance can be divided into two groups: technical, which, among others, encompasses reengineering, reverse engineering and restructuration; and management solutions, characterized by having quality assurance procedures, structured management, use of human resources specialized in maintenance, change documentation, etc. However, whereas every year new technical solutions are proposed, very little has been researched about management solutions. The consequences of this have been denounced by some authors: Pressman (1993), for example, affirms that there are rarely formal maintenance organizations, which implies that maintenance is done "willy nilly." Baxter and Pigdeon (1997) show that incomplete or out of date documentation is one of the main four problems of software maintenance. For Griswold and Notkin (1993), the successive software modifications make maintenance each time more expensive. For Pigoski, there is a lack of definition of the maintenance process which, furthermore, hampers the development and use of CASE tools for helping in its management. Also, excepting some works (Brooks, 1995; McConnell, 1997), the modeling of organizations is a neglected area of research in Software Engineering in general. Fugetta (1999) states that most of techniques and methods for supporting software processes should consider organizational aspects, as its structure or culture, which Pigoski (1997) particularizes for maintenance.

With this situation, our research group and Atos ODS (a multinational organization which provides maintenance services to big banking and industrial enterprises) decided to build up MANTEMA, a methodology for managing software maintenance which approaches the problem from the maximum possible number of view points:

1) Defining the complete process and structuring it in activities and tasks.
2) Providing references to techniques extracted from literature for executing the tasks, and proposing new techniques in those cases with little research activity.
3) Developing a structure for describing all people involved in maintenance process.
4) All this, constructed in such a manner that the defined process could be perfectly integrated with the rest of processes of the software life cycle.

This chapter presents the MANTEMA methdology for Software Maintenance. We start with a brief discussion of the work method we have followed to develop the methodology; then, the methodology is drawn, beginning with its macrostructure and going into a high detail level. Then, some techniques and metrics developed during MANTEMA construction are shown, as well as a list of some document templates generated in the maintenance process.

WORKING METHOD

During the development of MANTEMA, we followed the recommendations of Avison, Lau, Myer, and Nielson (1999) who say that "to make academic research relevant, researchers should try out their theories with practitioners in real situations and real organizations," and we decided to select Action Research as an appropriate method for working jointly with Atos ODS. Action Research is a qualitative research method. This kind of methods have received recently special attention (Seaman, 1999). According to McTaggart (1991), Action Research is "the way groups of people can organize the conditions under which they can learn from their own experiences and make this experience accessible to others". Basically, Action Research is an iterative research method which refines, after each cycle, the generated research products. In fact, Padak and Padak (1998) identify four cyclic phases in research projects carried out through Action Research (Figure 1).

As Wadsworth (1998) states, every refined solution provided after each cycle helps "to develop deeper understandings and more useful and more powerful theory about the matters we are researching, in order to produce new knowledge which can inform improved action or practice." (Wadsworth, 1998)

Figure 1. Cyclic process of action research.

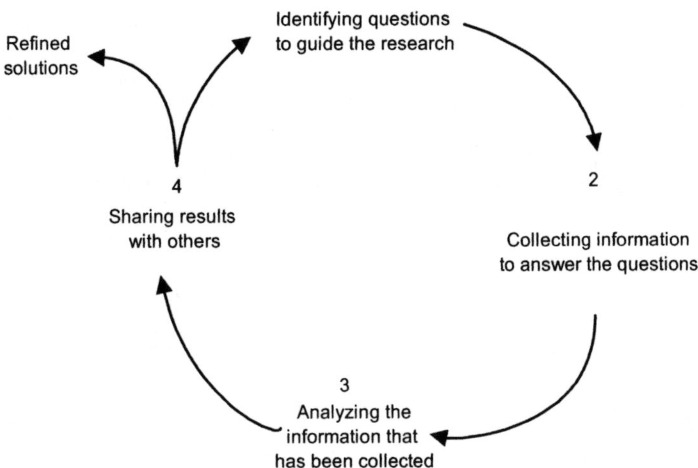

These cycles, coupled with the corresponding sharing of information and feedback for improving problem resolution, imply the participation of several parties. Four participant organizations are usually identified in Action Research projects:

1) The *researched object*, which is usually a problem that needs to be resolved and which, in this case, is the Maintenance Process and its improvement.
2) The *researcher/s*, the one who carries out the research, that has been constituted by our research group.
3) The *critical reference group*, which is the person/group who has the problem that the research is intended to resolve and which participates in the researching process. In our case, Atos ODS has played this role during the research process.
4) The *researched for*, that is, those who might benefit from better information about the situation, although they do not intervene directly in the research process. During MANTEMA construction, this fourth participant is constituted, for the first time, by all Atos ODS's customers and, for the second one, we hope that by all the software maintenance community.

The implementation of this work method has implied, during all the research process, a continuous feedback generated in the critical reference group and leaded to the researcher. In fact, the first meetings between both parties served to define the maintenance problem, explaining the Atos ODS's context to put the researcher in situation, and trying to draw the lines to the solution.

After this, several versions of the methodology have been proposed and put into practice. In this chapter, we present MANTEMA 2.0, generated as the product of the refinement of previous methodology releases applied in Spanish banking organizations.

RAISING MANTEMA METHODOLOGY

During the initial stages of MANTEMA construction, the researcher devoted a large part of the time to the research of previously defined methodologies for software maintenance. However, as we have noted, what we found was an absence of this kind of work that, on the other hand, convinced us to use more global frameworks as a methodological basis to start out.

For any maintenance organization to have software processes under control supposes undoubtable advantages; moreover, if such methodologies are built up from more global frameworks, as ISO/EC 12207 (ISO/IEC, 1995) or IEEE 1074 (IEEE, 1991), they can contribute to organizations with other kinds of added-values, such as getting quality certifications. This is also very interesting for supplier-services enterprises.

ISO/IEC 12207 and its Maintenance Process was selected as the foundation for constructing the methodology, since this International Standard defines all the software life cycle processes. This allows the integration of Maintenance with all of them and, as Pigoski (1997) says, this International Standard "will drive the world software trade and will impact on maintenance."

The direct use of the ISO/IEC 12207 Maintenance Process as a maintenance methodology has a series of drawbacks, some of which have been discussed in Polo, Piattini, Ruiz, and Calero (1999c) (for example: overlapping of certain activities and tasks produced by the insertion of a Development Process inside the Maintenance one; costs of adaptation; lack of specific consideration for outsourcing; lack of technical solutions). However, this is not really a design problem of ISO/IEC 12207 since it, as most of standards, specifies *what* to do, but not *how*. Furthermore, it is possible to adapt ISO/IEC 12207 to specific environments, projects, etc., through its Tailoring Process. It also serves to assure the compliance of the tailoring to the Standard: "Compliance is defined as the performance of all the processes, activities and tasks selected from this International Standard in the Tailoring Process for the software project" (ISO/IEC, 1995).

In Polo et al. (1999b), the obtaining of maintenance methodology from the application of the activities composing the Tailoring Process, mainly to the Maintenance one, is presented. The structure shown in Figure 2 can be extracted from the analysis of the ISO/IEC 12207 Maintenance Process, and constitutes an initial idea for starting out.

MANTEMA is a complete approach to the maintenance process, and tries to decrease the costs of all its activities. To get it, one of the firsts tasks is the clear definition of the set of activities to be executed along the maintenance process.

In this sense, it is useful to follow the recommendations proposed by some authors, in the sense of subjugating maintenance interventions to different sets of tasks, according to their maintenance type. Pressman (1993), for example, advises the classification of modification requests depending on whether or not they are due to an error, their degree of severity , etc. Also the IEEE Standard 1219 recommends the modification requests classification as corrective, perfective, preventive, and adaptive, and that they are integrated into sets that share the same design areas (IEEE, 1992). This idea is also mentioned in ISO/IEC 12207 but, as in IEEE 1219, activities to be followed depending on the type are not specified.

In MANTEMA, the following five types of maintenance are distinguished and precisely defined (Polo et al., 1999c):

1) Urgent corrective, when there is an error in the system which blocks it and must be corrected immediately.
2) Non-urgent corrective, when there is an error in the system which is not currently blocking it.
3) Perfective, when the addition of new functionalities is required.
4) Preventive, when some internal attributes of the software are going to be changed (maintainability, cyclomatic complexity, etc.), but with no change of either the functionality or the form of use of the system.
5) Adaptive, when the system is going to be adapted to a new operational environment.

This distinction allowed for the building of different technical guides for each type, which were progressively refined. However, in the final version of MANTEMA we have grouped the last four types into one, since the practical application of the methodology revealed that the treatment and ways of execution of these types of maintenance were very similar. Therefore, they have been grouped under an unique denomination, *planneable maintenance*, leaving in this manner the urgent corrective as *non-planneable maintenance.*

On the other hand, a method for establishing and ending outsourcing relationships has been incorporated into MANTEMA as a response to the growing importance that nowadays this area is obtaining in many sectors influenced by Information Technologies (Rao, Nam, & Chaudhury, 1996). Hoffman (1997) shows, for example, that 40% of the biggest companies in the United States have outsourced at least one of the major pieces of their operations.

Figure 2. Initial structure for constructing MANTEMA.

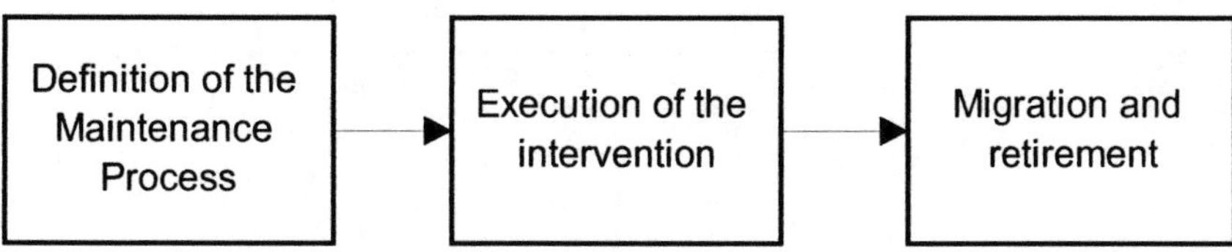

STRUCTURE OF MANTEMA

From the ideas exposed in the previous section, the basis of the maintenance process depicted in Figure 2 must be extended according to the different types of maintenance identified and the outsourcing activities. Figure 3 illustrates the results of these considerations.

Structure of Every Node

Every node in Figure 3 is composed of activities, composed in turn of a set of tasks. The first node includes activities to be executed at the beginning of the maintenance process and others devoted to the reception and classification of maintenance requests; all these are executed before any maintenance intervention. The last node groups the activities which must be executed after every intervention and at the end of the maintenance process. The central set contains the proper activities of every maintenance type.

For the complete definition of the methodology, it is important to define all the objects involved in the maintenance process. In MANTEMA, this is done at task level, in such manner that, for every task, we define:

1) Inputs, which are the needed objects for the correct execution of the task. These objects may be programs, documents, etc., and can be taken either from a previous task (for example, a modification request classified as perfective) or from the environment (a program to be improved).
2) Outputs: these are the objects generated by the execution of the task. Outputs can be leaded either to a following task (the classified modification request) or to the environment (an improved and fully tested program).
3) Techniques to be used during the task, some of which have been extracted from literature and others which have been proposed during MANTEMA construction.

Figure 3. Macrostructure of MANTEMA methodology.

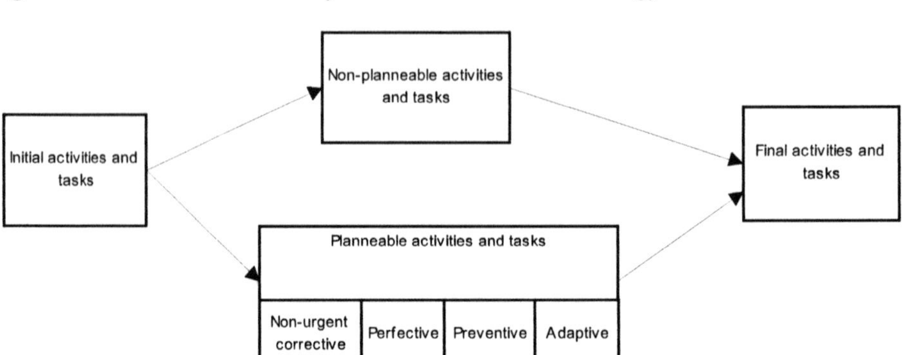

4) People responsible, who must belong to a set of profiles defined for the maintenance process.
5) Interfaces with other processes, as Configuration Management, in order to execute operations inside the organization's framework.

Then, the whole process can be seen as the concatenation of elements like those shown in Figure 4, which illustrates the schema of a task. As it is seen, some metrics must also be collected from the execution of every task, in order to have the process under control.

Roles in the Maintenance Process

The people responsible for every task must be identified during the Initial Study activity of the Common Initial Activities and Tasks. Every person involved in the maintenance process must belong to some of the organizations listed below and, moreover, his/her functions must correspond with those of some of the profiles mentioned in Polo et al. (1999d) and with further explanations in the chapter "Environment for Managing Software Maintenance Projects," in this book. On the other hand, depending on the case, two or even three organizations may coincide. The same can happen with profiles; two or more can be acted on by just one person. The main roles are:

1) Customer organization. This organization corresponds with the Acquirer defined in ISO/IEC 12207. We define it as the organization which owns the software and requires the maintenance service.
2) Maintainer, the organization which supplies the maintenance service.
3) User, the organization that uses the software.

Detailing Some Nodes of the Methodology

In this section, a detailed structure of the methodology is presented through a more complete explanation of activities and tasks.

Figure 4. Structure of a task.

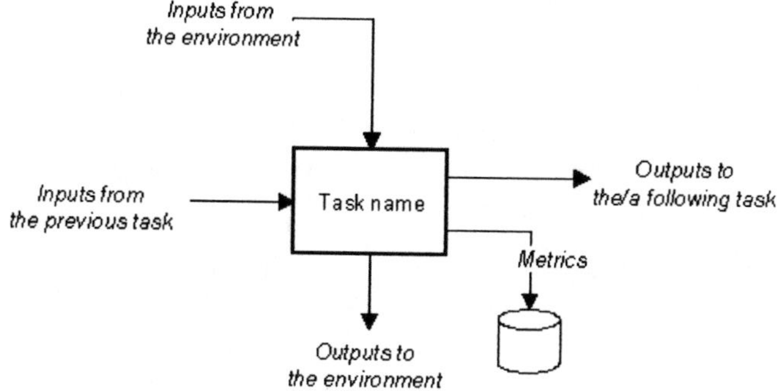

Common initial activities and tasks.

Activities and tasks of this initial node are executed from the moment when the future acquirer of the maintenance service contacts the outsourcing organization, until the reception and classification of modification requests is done. Some tasks and documents of this activity may be omitted if there is no outsourcing relationship.

The activities and tasks of this process are listed below, and a complete summary appears in Table 2.

Activity I0: Initial study.

This activity consists of the following tasks:

Task I0.1: Beginning and information recollection.

In this task, the Maintenance Organization must collect and analyze enough data to get an idea of the system to be maintained. This analysis is the main source of information used to elaborate the budget of the outsourcing, to schedule resources, etc. However, these factors also depend on what kind of maintenance services the Customer desires to outsource and on the response level at which the Maintenance Organization works.

In the subsequent signing of the contract, both parties will need to provide clear definitions in order to include all possible terms which are to be agreed upon. The firsts concepts that organizations must distinguish are the diverse types of maintenance previously mentioned, since it is normal that the customer does not outsource all the possible types (i.e., closed definitions of the different maintenance types to outsource must be provided).

Furthermore, different "service level agreements" will be put forward by the Maintenance Organization with different response levels, depending on the type of maintenance intervention (e.g. an urgent-corrective modification request should be executed in less time than others consisting of a misspelled word on an input screen).

As we can see in Table 2, the input for this task is a Request of Maintenance Service, and a Questionnaire with information about the software to be maintained must be produced by the Maintenance Organization. The contents of this questionnaire, which is generally filled in from a sample of the software, are detailed in Figure 5.

Task I0.2: Preparing the maintenance proposal.

In this task, the Maintenance Organization must prepare a pre-contract to be studied by the Customer. Obviously, the supplier must calculate estimates about maintenance costs, taking the data collected in the Questionnaire as a basis. Before to the presentation of the proposal, some internal reports must be done in order to estimate the terms of any future contract.

For this purpose, MANTEMA uses a checklist of risks to be considered by the Maintenance Organization. This checklist has been developed using some situational factors of Euromethod (1996) and from Atos ODS's own experience. It is important to bear in mind that, even though there is a lot of good research analyzing the problem of risks once the maintenance process has started (Schneidewind, 1997; Sherer, 1997), consideration of risks for outsourcing "has received little systematic academic attention" (Willcocks & Lacity, 1999) overall from the Supplier Organization's point of view. Techniques for cost estimations are

Figure 5. Contents of the questionnaire.

A) Organization identification (financial data, personnel, etc.) and authorized speakers
B) Hardware and software environment
  B.1) Hardware
    B.1.1.- Mainframes (mark and model, operating system, etc.)
    B.1.2.- Secondary hardware
      1.2.1.- Terminals
      1.2.2.- Workstations (indicating operating system)
      1.2.3.- Network type
      1.2.4.-Connection type with the main hardware
  B.2) - Software environment
    B.2.1.- Development environment
    B.2.2.- File system
    B.2.3.- Databases
    B.2.4.- Batchs JCL
C) Development and maintenance organization
  C.1.- Development methodologies and techniques
  C.2.- Use of denomination and programming standards
  C.3.- Standards and techniques of quality management
  C.4.- Project management procedures
  C5.- Operation procedures
  C.6.- Auditing procedures
  C.7.- Problem resolution procedures
  C.8.- Documentation procedures
  C.9.- Other procedures or standards
D) Applications (for every one)
  D.1.- Identification (name and date of putting into production)
  D.2.- Organizational unit responsible/user
  D.3.- Related applications
  D.4.- Batch programs (languages, size, year of writing and number of interventions)
  D.5.- Online programs (languages, size, number of modules, year of writing, and number of interventions)
  D.6.- Reports (number, year of writing, and number of interventions)
  D.7.- Screens (number, year of writing, and number of interventions)

Figure 6. Contents of the maintenance proposal.

1) Introduction.
2) General results of the analysis applied to the applications.
3) Service proposal:
   3.1 Technical report defining: goals, bounds, bindings, responsibilities, and contractual parameters.
   3.2 For every application inside the (future) maintenance contract, maintenance types and their corresponding service indicators must be set.
   3.3 Proposal of contract (that will have the same format as the Definitive contract).
4) Economic offer.

a possible exception at this point; however, they do not consider risks in such estimations (although maybe the table of weight assignation of Albrecht's, 1979, function-point could be considered as a light way of hefting risks) or, if they do, they are given over to an expert's judgement (Briand, El-Eman, & Bomarius, 1998). In Polo, Piattini, and Ruiz (2002), the complete method for identifying and estimating risks in Software Maintenance projects is presented in detail.

These results are used to covenant the values of the Service Level Agreements between the two involved parties when there is an outsourcing relationship. Later in this chapter, the use of these agreements for planning the needed resources for urgent-corrective is shown.

With all these data, a Maintenance Proposal containing at least the following contents can be filled in; Figure 6 provides the minimal content of such a proposal.

Task I0.3: Contract.

Once the Maintenance Proposal has been analyzed and discussed by both parties, a contract detailing the terms as listed in Figure 7 is signed.

Task I0.4: Planning customer/supplier relations.

In this task, a calendar detailing the future control meetings must be set up. As a good these meetings will make sure that the customer knows the status of its applications portfolio, although they can also be understood as milestones for controlling the work of the maintenance organization.

Besides, people responsible of each organization (Customer, User and Maintenance Organization) must be identified, determining who will represent each one of the profiles mentioned in Polo et al. (1999d).

Activity I1: Process implementation.

Although in this task the customer and the maintenance organizations have already signed the maintenance service contract, the maintenance organization does

Figure 7. Contents of the maintenance contract.

1) Identification of the parts
2) Object of the contract (*the customer relies on the provision of a set of presentations on the field of the computer services that will make up the maintenance of the software registered in the context of the technical proposal*).
3) Characteristic of the service:
   3.1. Inventory of the objects software to maintain.
   3.2. Initial state of the software.
   3.3. Conditions of organization of the work of the provider and of the client.
   3.4. Formalization conditions of the maintenance intervention.
   3.5. Adaptation of the quality assurance plan to the customer.
   3.6. Correction of anomalies.
   3.7. Maintenance of the competitive level of the application software and of the standard software.
   3.8. Documentation facilitated to the attached presentation.
   3.9. Assistance modality to the final users.
4) Obligations of the customer.
5) Obligations of the provider.
6) Clauses of exclusion.

not yet have the responsibility of executing maintenance interventions. Such responsibility will arrive at the end of this activity, when the tasks composing this activity have been fully executed. The following are the tasks related to this activity:

Task I1.0: Knowledge acquisition.

In this task, the maintenance team studies the existing documentation about the software to be maintained, including source code, databases schemas, etc. Interviews with users and the paying attention of how the current maintenance organization works are also required. During this task, the customer (or its current maintenance organization) must resolve the maintenance requests, since the new maintenance organization is now learning the structure and operation modes of the software.

This task takes approximately one or two months, and during this time the new maintainer will probably not have modified one line of code. At the end of this task, the new maintenance organization must provide the customer with updated documentation about the software. However, this task may be partially omitted if the software has not been maintained previously.

Task I1.1: Developing plans.

The Maintenance Team develops the maintenance plans and builds up a Technical Summary of the software. The Technical Summary must collect information about project goals, main functions, technical and management constraints, a summary of the data obtained in the questionnaires, historical data, estimating techniques, control mechanisms, and, for every application and database, the following metrics:

- Number of modules.
- Number of lines of code (according to some standard definition).
- Mean cyclomatic complexity of every module.
- Number of global variables and routines (externally visible) of every module (this will be useful for identifying fault-prone modules).
- Metrics of databases.

Task I1.2: Defining modification requests procedures.

The Maintenance Organization generates document templates for modification requests and, with the Customer, defines the procedures for their presentation.

Task I1.3: Implementing software configuration management process.

As we can see in Table 1, in this task an interface with the Configuration Management process is established in order to tailor this process (if it exists) to the current system and context, or for defining a new one.

Task I1.4: Preparing tests environments.

The Maintenance Team must prepare copies of the software product for future intervention tests on the software.

From this moment, the new maintenance organization will be responsible for the execution of all the maintenance interventions, from receipt until passing to production environment.

Activity I2: Study of the modification request.

Task I2.1: Reception of the Modification Request

The Maintainer receives a Modification Request from the Customer. This receipt must be registered.

Task I2.2: Decision about the type of maintenance

From the Modification Request received and registered in the previous task, the Maintainer decides which of the five maintenance types must be applied. After this, the Maintainer puts the Modification Request into the queue, maybe passing it by other older Modification Requests.

Table 2. Common initial activities and tasks.

Activities →	Initial study				Process implementation					Study of the modification request	
Tasks →	I0.1 Beginning and information recollection	I0.2 Preparing maintenance proposal	I0.3 Contract	I0.4 Planning customer/ supplier relations	I1.0 Knowledge acquisition	I1.1 Developing plans	I1.2 Defining modification request procedures	I1.3 Implementing Configuration Management process	I1.4 Preparing test environments	I2.1 Receipt of the modification request	I2.2 Decision about the type of intervention
Inputs	Request of Maintenance service	Questionnaire Interviews	Maintenance proposal	Maintenance contract	Software product in operation	Software product in an operation Maintenance plan	Maintenance plan	Software product in operation Maintenance plan	Software product in operation	Modification request	Received modification request
Outputs	Questionnair e	Maintenance proposal	Maintenance contract	Customer/ supplier relations planning (responsibiliti es, meetings calendar, etc.)	Software product in operation yet new Documents about such software		Document models and forms Regulatory ways	Configuration Management Process	Copies of the software elements in operation	Received modification request	Report about type of maintenance, criticise, etc. Decision about the type of maintenance
Techniques	Interview		Estimation		Observation Cross references				Configuration management		
Responsible	Customer Maintainer	Maintainer	Maintainer Customer	Maintainer Customer	Maintainer	Maintainer	Maintainer	Maintainer	Maintainer	Customer Maintainer	Maintainer
Interfaces with other processes				Tailoring				Configuration Management	Quality assurance		

Non-planneable (urgent corrective) activities and tasks.

This case applies when there is an error which seriously paralyzes the normal system or organization operation and must be corrected quickly.

Below we describe briefly the activities and tasks of this type of maintenance, which is summarized in Table 3.

Activity UC1: Error analysis.

This activity consists of the following task:

Task UC1.1: Investigating and analyzing causes.

The Maintainer analyzes the Modification Request, verifies the problem (perhaps with User collaboration), reproduces it, and studies different alternatives for implementing the change. Furthermore, a list of software elements to be changed (modules, routines, documentation, etc.) must be developed.

The following metrics must be collected in this task:

- Time dedicated to the task.
- Number of affected function points.
- Error origin (Briand et al., 1998).
- Error cause (Briand et al., 1998).

Activity UC2: Urgent-corrective intervention.

This activity consists of the following tasks:

Task UC2.1: Making corrective actions.

Maintainer executes the needed corrective actions to correct the detected problem. The following metrics must be collected in this task:

- Time dedicated to the task.
- Number of altered modules.
- Number of altered routines.
- Number of added lines of code (according to some standard definition).
- Number of modified lines of code (according to some standard definition).
- Number of deleted lines of code (according to some standard definition).
- Mean cyclomatic complexity of every altered module.
- Number of global variables and routines (externally visible) of every module.
- Metrics of databases.

The continuous saving of product metrics help to the customer understand values of preventive maintenance that the Maintainer needs to execute during interventions. This is especially interesting when there is an outsourcing relationship and the external organization gets a contract which commits to (for example) a progressively decreasing cyclomatic complexity of modules while it corrects errors.

Task UC2.2: Complimenting documentation.

The maintainer must document the changes in the Corrective Actions Realized document. The following metric must be collected in this task:

- Time dedicated to the task.

Task UC2.3: Verification of the modification.

The Maintainer must test the correction of all the changes. These tests are documented in the Unitary Tests document.

The following metrics must be collected in this task:

- Time dedicated to the task.
- Number of detected high-gravity errors in tests.
- Number of mean-gravity errors detected in tests.
- Number of little-gravity errors detected in tests.

Activity UC3: Intervention closing.

This activity consists of the following task:

Task UC3.1: Putting the software product into production environment.

The changed software is put into the production environment to be used by the users. An interface with the Configuration Management Process is set to guarantee the correction of this task.

The following metrics must be collected in this task:

- Time dedicated to the task.

Planneable maintenance.

As was previously mentioned, under this denomination we have grouped the non-urgent corrective, perfective, preventive, and adaptive types of maintenance, because they share a large portion of the activities and tasks. However, every one of them has some certain particularities which distinguish it from the others. The different task ways for every one of these types of maintenance is shown in Figure 8, and we detail them in tables 4 and 5.

Common final activities and tasks.

The following five activities compose this last node of the graph shown in Figure 3.

Activity CF1: Intervention registering.

This activity comprises of the following task:

Task CF1.1: Intervention registering.

The intervention is registered according to organization procedures.

Table 3. Urgent corrective activities and tasks.

Activities→	Error analysis	Corrective intervention			Intervention closing
Tasks→	UC1.1 Investigating and analyzing causes	UC2.1 Making corrective actions	UC2.2 Complimenting documentation	UC2.3 Verifying the modifications	UC3.1 Putting the software product into production environment
Inputs	Software product in operation Urgent error Modification request	Software to be corrected	Old software (with errors) New software (without visible errors)	Corrected software Unitary test cases Integration test cases	Corrected software
Outputs	Software to be corrected	Corrected software	Documentation with corrective actions	Assurance of the correction of the error	Corrected software in operation
Techniques	Source code analysis Error reproduction Documentation studying	Codification	Redocumentation	Test techniques	
Metrics	Time dedicated to the task. Number of affected function points. Error origin. Error cause.	Time dedicated to the task. Number of altered modules. Number of altered routines. Number of added lines of code. Number of modified lines of code. Number of deleted lines of code. Mean cyclomatic complexity of every altered module. Number of global variables and routines of every module. Metrics of databases.	Time dedicated to the task.	Time dedicated to the task. Number of detected high-gravity errors in tests. Number of mean-gravity errors detected in tests. Number of little-gravity errors detected in tests.	Time dedicated to the task.
Responsible	Maintainer	Maintainer	Maintainer	Maintainer	Maintainer
Interfaces with other processes				Verification	Configuration Management

Table 4. Planneable maintenance (continued).

Activity→	Modification Request Analysis			Intervention and tests (continues)			
Applicable type of maintenance	CP/P/A	CP/P	CP	A	A	CP/P/A	CP/P/A
Tasks→	P1.1 MR assessment	P1.2 Documenting possible solutions	P1.3 Selecting alternative	P2.1 Planning calendar	P2.2 Copying software product	P2.3 Executing intervention	P2.4 Unitary tests
Inputs	Software product in operation Modification request	Software product in operation MR in the waiting queue	Software product in operation Implementation alternatives (DOC10)	List of software elements and properties to improve (DOC8) Project documentation	Software product in operation	CP: Software product in operation Error diagnostic (DOC9) Selected alternative P: Software product in operation List of software elements and properties to improve (DOC8) A: Copy of the software product	Modified software product Document of intervention (DOC7/11/13) Modification request)
Outputs	MR in the waiting queue Intervention calendar A: Schedule estimation Resources disposability	CP: Error diagnostic and possible solutions (DOC9) Implementation alternatives (DOC10) Product measures (DOC16a) P: List of software elements and properties to improve (DOC12) Product measures (DOC16a)	Selected alternative (full DOC9)	Intervention calendar	Copy of the software product	CP: Corrected software product Document of intervention (DOC7/DOC11) P: Modified software product Document of intervention (DOC13) A: Adapted copy	Unitarily tested software product Document of unitary tests (DOC8
Techniques	Portfolio Analysis Project management	Source code analysis Project documentation analysis	Query to the historical DB	Project management		Codification Redocumentation	Test techniques
Metrics	Time dedicated to the task	Time dedicated to the task Number of affected FP Error origin and cause	Time dedicated to the task	Time dedicated to the task	Time dedicated to the task	*See task UC2.1 in* ***Error! Reference source not found.***	*See task UC2.3 in* ***Error! Reference source not found.***
Responsible	Maintainer	Maintainer	Maintainer	Maintainer	Maintainer	Maintainer	Maintainer
Int. processes		Quality assurance		Config. Manag.	Config. Manag.	Quality assurance	Quality assurance

Table 5. Planneable maintenance (continued).

Activity→	Modification Request Analysis			Intervention and tests (continues)			
Applicable type of maintenance	CP/P/A	CP/P	CP	A	A	CP/P/A	CP/P/A
Tasks→	P1.1 MR assessment	P1.2 Documenting possible solutions	P1.3 Selecting alternative	P2.1 Planning calendar	P2.2 Copying software product	P2.3 Executing intervention	P2.4 Unitary tests
Inputs	Software product in operation Modification request	Software product in operation MR in the waiting queue	Software product in operation Implementation alternatives (DOC10)	List of software elements and properties to improve (DOC8) Project documentation	Software product in operation	CP: Software product in operation Error diagnostic (DOC9) Selected alternative P: Software product in operation List of software elements and properties to improve (DOC8) A: Copy of the software product	Modified software product Document of intervention (DOC7/11/13) Modification request)
Outputs	MR in the waiting queue Intervention calendar A: Schedule estimation Resources disposability	CP: Error diagnostic and possible solutions (DOC9) Implementation alternatives (DOC10) Product measures (DOC16a) P: List of software elements and properties to improve (DOC12) Product measures (DOC16a)	Selected alternative (full DOC9)	Intervention calendar	Copy of the software product	CP: Corrected software product Document of intervention (DOC7/DOC11) P: Modified software product Document of intervention (DOC13) A: Adapted copy	Unitarily tested software product Document of unitary tests (DOC8
Techniques	Portfolio Analysis Project management	Source code analysis Project documentation analysis	Query to the historical DB	Project management		Codification Redocumentation	Test techniques
Metrics	Time dedicated to the task	Time dedicated to the task Number of affected FP Error origin and cause	Time dedicated to the task	Time dedicated to the task	Time dedicated to the task	*See task UC2.1 in* ***Error! Reference source not found.***	*See task UC2.3 in* ***Error! Reference source not found.***
Responsible	Maintainer	Maintainer	Maintainer	Maintainer	Maintainer	Maintainer	Maintainer
Int. processes		Quality assurance		Config. Manag.	Config. Manag.	Quality assurance	Quality assurance

Figure 8. Structure of the planneable maintenance.

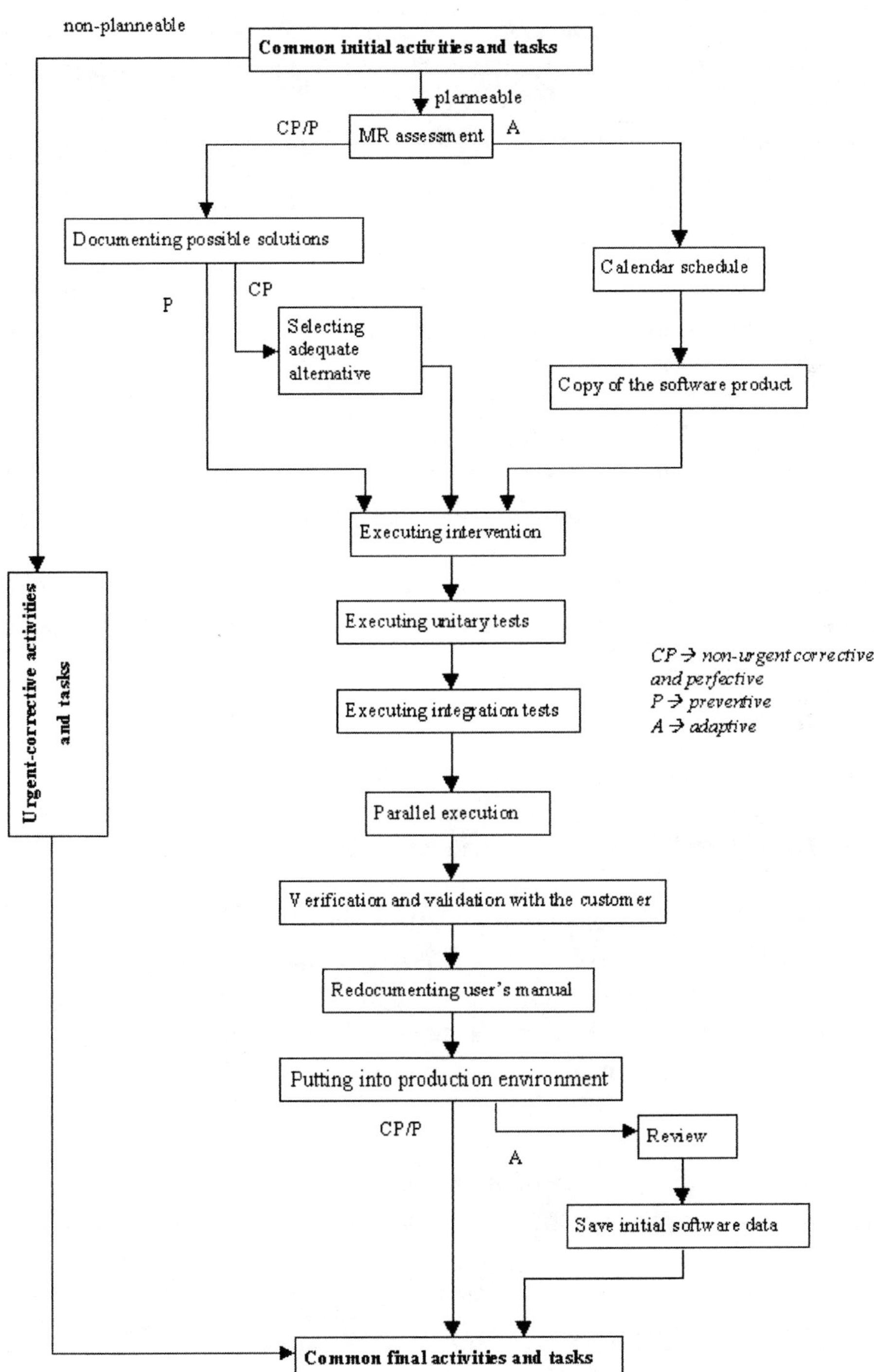

Activity CF2: Historic database updating.

This activity consists of the following tasks:

Task CF2.1: Collecting intervention information.

The Maintainer collects information about time and resources spent on the maintenance intervention which has ended, and also about metrics of the product before and after the intervention. There is a possibility that this task has been executed with the metric recollection during the intervention.

Task CF2.2: Historic database updating.

Historic data collected in the previous task are included in the historic database. Maybe this task has been executed during the process if a tool, like MANTOOL, has been used for managing maintenance.

Activities CF3 and CF4: Migration and Retirement.

We omit descriptions of these activities since they have been directly imported from ISO/IEC 12207. The execution of these activities will not be always needed.

Activity CF5: End of outsourcing.

Tasks in this activity will be executed when there is an outsourcing relationship between two organizations.

Task F5.1: Inventory and documentation delivery.

Depending on the contract, the Maintenance Organization must deliver to the Customer all the generated and modified software products during the period in which it has been responsible for the maintenance process.

Task F5.2: Training and devolution of experience.

This task is the inverse of task I1.0 (*Knowledge acquisition*), during which the Maintenance Team learned the characteristics of the system by watching the work mode of the maintainer that, in that moment, had the Customer. Now it is the Maintenance Organization who must train the new maintenance personnel, allowing maintenance with mixed teams.

Task F5.3: Definitive delivery of the service.

The Maintenance Organization finishes definitively the benefit of the service to the Customer.

TECHNIQUES, METRICS, AND DOCUMENTS

In this section, an introduction to some new techniques and metrics to be used during the process is done.

Planning the Non-Planneable Maintenance

Kung and Hsu (1998) have modeled a typical lifecycle for maintenance, including corrective. For example, we can observe how most requests of this type of maintenance are concentrated in the later moments of every software release. Also Basili et al. (1996) and Schneidewind (1998) have shown similar results. Calzolari, Tonella, and Antoniol (1998) have proposed a method which for predicting the arrival of maintenance requests, including corrective. The ANGEL tool of Shepper, Schofield, and Kitchenham (1996), which predicts effort using analogy has been used by Niessink and Van Vliet (1997) in maintenance projects.

In MANTEMA, we may use any predictive model of maintenance request arrival, together with economical parameters of the project for determining the quantity of resources to plan for error corrections with no economical lose (Polo et al., 2002). Nowadays, there is in Europe an excess of demand for skilled personnel in Information Technologies: 500,000 persons in 1998, which will be 2.4 million in 2002 (Bourke, 1999). So, this method is very useful to those organizations which have a limited number of resources to be dedicated to several projects, every one with different economical parameters.

Process Metrics

In this section, some metrics to maintain the control of the process are shown:

a) Number of received MRs (NRMR)
b) Ratio of rejected/received MRs
c) Ratio of mutually incompatible MRs
d) Number of urgent-corrective MR/NRMR
e) Number of non-urgent corrective MR/NRMR
f) Number of perfective MR/NRMR
g) Number of preventive MR/NRMR
h) Number of adaptive MR/NRMR
i) Number of MR due to legal changes/NRMR
j) Number of MR due to business evolution/NRMR
k) Number of MR due to business rules change/NRMR
l) Number of replanned MR per period and area, period and NRMR, period and functional domain
m) Service indicators:
   - Response time
   - Planning respect
n) Flexibility with no replanning: number of persons-month we can dispose with no replanning nor new resources allocation
o) Number of persons for each project and month
p) Medium degree of replanning: number of hours replanned in each period.

q) Medium response time in urgent MR
r) Medium number of monthly MR
s) Percentage of anomalies whose responsibility is of the Maintenance Team
t) Number of hours, function-points, etc., replanned /Total hours, function-points, etc., initially planned for each period
u) Tendencies

Documents

During the maintenance process, a large quantity of documentation is generated. MANTEMA provides standard templates for documents as shown in Figure 9 (some of which have been presented in Figures 5, 6 and 7.)

Automatic Support

In general, there are many tools for helping in the maintenance process, although most of them were initially designed for other goals (Quang, 1993). However, nearly all of these are vertical tools, in the sense that they only help in some of the maintenance tasks (cost estimation, reengineering, reverse engineering, restructuration) and there are few horizontal tools suitable of being used along the entire process; moreover, the use of horizontal tools is limited to certain bands of the maintenance process, such as configuration management (Pigoski, 1997).

Figure 9. List of document templates in MANTEMA.

DOC1 - Initial questionnaire.
DOC2 - Maintenance proposal.
DOC3 - Maintenance contract.
DOC4 - Risks factors table.
DOC5 - Technical summary.
DOC6 - Modification request.
DOC7 - Executed corrective actions.
DOC8 - Executed unitary tests.
DOC9 - Diagnostic and possible solutions.
DOC10 - Implementation alternatives.
DOC11 - Executed perfective actions.
DOC12 - List of software elements and properties to improve.
DOC13 - Executed preventive actions.
DOC14. Migration plan.
DOC15. Note of future migration.
DOC16. Product measures.
DOC17. Plan of maintenance for the period.

The rigid definition of the maintenance process done by MANTEMA, structuring for its activities and tasks for every one of the particular types of maintenance, and providing all the inputs, outputs, metrics, etc., for every task, thus allows the easy translation of the process into an automatic environment. Moreover, one of the biggest problems of the first applications of MANTEMA related by practitioners was the lack of a tool for using the methodology. Therefore, the implementation of MANTOOL, a tool for managing the maintenance process according to MANTEMA, was a must for the project.

MANTOOL is presented in detail in Polo et al. (2001) and is a part of the MANTIS environment, which is presented in the chapter "Environment for Managing Software Maintenance Projects," also contained in this book.

CONCLUSIONS

In this chapter we have presented MANTEMA, a methodology for supporting the maintenance process. We believe that this methodology covers a lack of Software Engineering, providing a complete guide for carrying out the process through the rigorous definition of different types of maintenance, all of them detailed at task level.

Such process architecture allows the easy translation of the methodology to an automatic environment, which facilitates the application of the methodology and the control of the process.

REFERENCES

Albretch, A. J. (1979). Measuring application development productivity. *Proceedings of the IBM Application Development Symposium.* Monterey, Canada.

Avison, D., Lau, F., Myers, M., & Nielsen, A. (1999). Action research. *Communications of the ACM,* 42(1), 94-97.

Basili, V., Briand, L., Condon, S., Kim, Y., Melo, W., & Valett, J.D. (1996). Understanding and Predicting the Process of Software Maintenance Releases. *Proceedings of the International Conference on Software Engineering. IEEE.*

Baxter, I.D. & Pidgeon, W.D. (1997). Software Change Through Design Maintenance, *Proceedings of the International Conference on Software Engineering,* 250-259. IEEE Computer Society, Los Alamitos, California.

Bourke, T. (1999). *Seven majors ICT companies join the European Commission to work towards closing the skills gap in Europe.* Conclusions of the Conference on "New Partnerships to Close Europe's Information and Communication Technologies." Available online: http://www.career-space.com/whats_new/press_rel.doc. Accessed January 2, 2000.

Brereton, P., Budgen, D., & Hamilton, G. (1999). Hypertext: The next maintenance mountain. *Computer,* 31(12), 49-55.

Briand, L., Kim, Y., Melo, W., Seaman, C., & Basili, V.R. (1998). Q-MOPP: Qualitative evaluation of maintenance organizations, processes and products. *Software Maintenance: Research and Practice*, 10, 249-278.

Briand, L.C., El Emam, K., & Bomarius, F. (1998). COBRA: A hybrid method for software cost estimation, benchmarking and risk assessment. *Proceedings of the 20th International Conference on Software Engineering.* Washington D.C: IEEE Computer Society Press.

Brooks, F. P. (1995) The mythical man-month. Essays on software engineering Anniversary Edition. Reading, MA: Addison-Wesley.

Calzolari, F., Tonella, P., & Antoniol, G. (1998). Modelling maintenance effort by means of dynamic systems. *Proceedings of the Third European Conference on Software Maintenance and Reengineering.* Amsterdam (The Netherlands), IEEE Computer Society, Los Alamitos, CA, USA.

De Vogel, M. (1999). Outsourcing and metrics. *Proceedings of the 2nd European Measurement Conference,* FESMA'99. Federation of European Software Metrics Association/Technologisch Instituut.

Euromethod Version 1. 1996. Euromethod Project.

Frazer, A. (1992). Reverse engineering-Hype, Hope or Here? In P.A.V. Hall (ed.), Software reuse and reverse engineering in practice. p. 209-243. Chapman & Hall.

Fuggetta, A. (1999). Rethinking the models of software engineering research. *The Journal of Systems and Software,* (47), 133-138.

Griswold, W.G. & Notkin, D. (1993). Automated assistance for program restructuring. *ACM Transactions on Software Engineering and Methodology*, 2(3), 228-269.

Hoffman, T. (1997). Users say move quickly when outsourcing your personnel. *Computer World*, March, p.77.

IEEE Std. 1074-1991 (1991). Standard for Developing Software Life Cycle Processes. IEEE Computer Society Press.

IEEE Std. 1219-1992 (1992). *Standard for Software Maintenance.* New York: IEEE Computer Society Press.

ISO-International Standard Organization (1995). ISO/IEC 12207. *Information Technology Software Life Cycle Processes.*

Jahnke, J.H. & Wadsack, J. (1999). Integration of Analysis and Redesign Activities in Information System Reengineering. *Proceedings of the Third European Conference on Software Maintenance and Reengineering,* Amsterdam. Los Alamitos, CA: IEEE Computer Society.

Jones, C. (1994). *Assessment and Control of Software Risks*. New York: McGraw-Hill.

Jørgensen, M. (1995). Experience with the accuracy of software maintenance task effort prediction models. *IEEE Transactions on Software Engineering*, 21(8), 674-681.

Karolak, D.W. (1996). *Software Engineering Risk Management.* Los Alamitos CA: IEEE Computer Society Press.

Kung, H.-J. & Hsu, Ch. (1998). Software maintenance life cycle model. *Proceedings of the International Conference on Software Maintenance.* Los Alamitos, CA, IEEE Computer Society.

Lehman, M, M. (1980). Programs, life cycles and laws of software evolution. *Proceedings of the IEEE,* 19, 1060-1076.

Lehman, M.M., Perry, D.E., & Ramil, J.F. (1998). Implications of evolution metrics on software maintenance. *Proceedings of the International Conference on Software Maintenance.* Los Alamitos, CA: IEEE Computer Society.

Lientz, B.P. & Swanson, E.F. (1980). Software maintenance management. Ed. Reading, MA: Addison-Wesley.

McConnell, S. (1997). Desarrollo y Gestión de Proyectos Informáticos. New York: McGraw Hill-Microsoft Press.

McTaggart, R. (1991). Principles of participatory action research. *Adult Education Quarterly*, 41(3).

Niessink, F. & Van Vliet, H. (1997). Predicting Maintenance Effort with Function Points. *Proceedings of the International Conference on Software Maintenance.* Los Alamitos, CA: IEEE Computer Society.

Padak, N. & Padak, G. (1998). Guidelines for planning action research projects. Available online: http://archon.educ.kent.edu/Oasis/Pubs/0200-08.html.

Piattini, M., Calero, C., Polo, M., & Ruiz, F. (1998). Maintainability in object-relational databases. *Proceedings of the European Software Measurement Conference FESMA 98,* Antwerp, May 6-8, 223-230.

Pigoski, T. M. (1997). *Practical Software Maintenance. Best Practices for Managing Your Investment.* New York: John Wiley & Sons.

Polo, M., Piattini, M, & Ruiz, F. (2001). Mantool: A tool for supporting the software maintenance process. *Journal of Software Maintenance and Evolution: Research and Practice,* 13(2), 77–95.

Polo, M., Piattini, M., & Ruiz, F. (2002). Integrating outsourcing in the maintenance process. *Information Technology and Management,* 3(3), 24–269.

Polo, M., Piattini, M., Ruiz, F., & Calero, C. (1999a). MANTEMA: A software maintenance methodology based on the ISO/IEC 12207 standard. *Proceedings of the 4th International Software Engineering Standards Symposium and Forum,* Curitiba (Brasil). Los Alamitos, CA: IEEE Computer Society.

Polo, M., Piattini, M., Ruiz, F., & Calero, C. (1999b). Using the ISO/IEC 12207 tailoring process for defining a maintenance process. *Proceedings of the First*

IEEE Conference on Standardisation and Innovation in Information Technology (SIIT '99). Aachen, Germany.

Polo, M., Piattini, M., Ruiz, F., & Calero, C. (1999c). MANTEMA: A complete rigorous methodology for supporting maintenance based on the ISO/IEC 12207 Standard. *Proceedings of the Third European Conference on Software Maintenance and Reengineering*. Los Alamitos, CA: IEEE Computer Society.

Polo, M., Piattini, M., Ruiz, F., & Calero, C. (1999d). Roles in the maintenance process. *Software Engineering Notes*, 24(4), 84-86.

Pressman, Roger S. (1993). *Software Engineering: A Practitioner's Approach* (Third edition). McGraw-Hill.

Project Management Institute (1996). *A guide to the project management body of knowledge*. Newtown Squares, PA: Project Management Institute.

Quang, P. T., (1993). *Réussir la Mise en Place du Génie Logiciel et de l'Assurance Qualité*. Ed. Eyrolles. France.

Rao, H.R., Nam, K., & Chaudhury, A. (1996). Information Systems outsourcing. *Communications of the ACM*, 39(7), 27-28.

Schach, S.R. (1990). *Software engineering*. Irwin & Aksen, USA.

Schneidewind, N.F. (1997). Measuring and evaluating maintenance process using reliability, risk and test metrics. *Proceedings of the 19th International Conference on Software Maintenance*, Los Alamitos CA: IEEE Computer Society Press.

Schneidewind, N.F. (1998). How to evaluate legacy systems maintenance. *IEEE Software*, July/August, 34-42.

Seaman, C.B. (1999). Qualitative methods in empirical studies of software engineering. *IEEE Transactions on Software Engineering*, 25(4), 557-572.

Shepper, M., Schofield, C., & Kitchenham, B. (1996). Effort estimation using analogy. *Proceedings of the 18th International Conference on Software Maintenance*. Los Alamitos, CA: IEEE Computer Society Press.

Sherer, S.A. (1997). Using risk analysis to manage software maintenance. *Journal on Software Maintenance: Research and Practice*, 9, 345-364.

Singer, J. (1998). Practices of Software Maintenance. *Proceedings of the International Conference on Software Maintenance*.

Sneed, H.M. & Foshag, O. (1998). Measuring legacy databases structures. *Proceedings of the European Conference on Software Measurement*. 199-211.

Wadsworth, Y. What is participatory action research? *Action Research International*, Paper 2. Available online: http://www.scu.edu.au/schools/sawd/ari/ari-wadsworth.html. Accessed January 2, 2000.

Willcocks, L.P. & Lacity, M.C. (1999). IT outsourcing in insurance services: Risks, creative contracting and business advantage. *Information Systems Journal*, 9(3), 163-180.

Chapter X

Environment for Managing Software Maintenance Projects

Francisco Ruiz, Félix García, Mario Piattini, Macario Polo
Escuela Superior de Informatica,
Universidad de Castilla - La Mancha, Spain

A Software Engineering Environment (SEE) is quite useful in order to manage the complexity of SM projects, since it can provide the needed services. Of the different aspects to highlight in these environments, in this chapter we put our main attention on those that are more directly related to the goal of helping in the management of SM complexity: to approach the SMP from a wide perspective of business processes to integrate technological and management aspects; to define a Process-centered Software Engineering Environment (PSEE); and to use a multilevel conceptual architecture based on standards like MOF (Meta-Object Facility). The MANTIS proposal of integral environment for the management of SM projects is also presented, and the main components of this environment are commented: conceptual tools (multilevel architecture, ontologies, software processes models and metamodels); methodological tools (methodology, and interfaces with organizational and managerial processes) and technical tools (horizontal and vertical software tools, repository, and interaction with process enactment software tools).

In the area of Software Maintenance (SM), there are still a number of matters to study and research (Bennett & Rajlich, 2000). One of the most important is the development of tools and environments to support methodologies and to facilitate the reuse of processes (Harrison, Ossher, & Tarr, 2000). A Software Engineering Environment (SEE) is quite useful to manage the complexity of SM projects, since it can provide the needed services. The SEE must be capable of managing data and metadata of the different production processes – in our case, the Software Maintenance Process (SMP) – at different detail and abstraction levels. The SEE should be based, for this purpose, upon a conceptual multilevel architecture, allowing all the information of processes to be shared among all available tools. This last need is satisfied using a repository manager that saves data and metadata of processes using an open and portable format.

In this chapter, a conceptual multilevel architecture is presented, making the integration of all available tools for managing SM projects possible, in precisely an integrated environment.

The fact is that such a SEE is a help to approach the inherent complexity of the SMP from a broader perspective than the merely technological one.

A SEE must satisfy several requirements to reach the aforementioned general goal. The two most meaningful requirements are the following: it must be process-oriented, and it must permit work with different models and metamodels of the software processes involved in the SM projects.

Of the different aspects to highlight in these environments, in this chapter we put our main attention on those that are more directly related to the goal of helping in the management of complexity. The first section presents a proposal to approach the SMP from a wide perspective of business processes, integrating technological and management aspects. The use of a Process-sensitive Software Engineering Environment (PSEE) to reach such goal is justified in the second section, and its architecture is presented.

The importance of using a multilevel conceptual architecture are justified in the third section and how to apply the Meta-Object Facility (MOF) Standard to the SM is commented. The MANTIS proposal of integral environment for the management of SM projects is presented in the following section. Lastly, the main components of MANTIS are commented in the remaining sections: conceptual tools (multilevel architecture, involved processes, ontologies, and metamodels); methodological tools (methodology and interfaces with organizational and managerial processes) and technical tools (horizontal and vertical software tools, repository, and interaction with process enactment software tools).

SOFTWARE MAINTENANCE AS A BUSINESS PROCESS

In recent years, everything that occurs once a software product has been delivered to users and clients has been receiving much more attention because of the significant economic importance that it has on the information technology industry. Proof of this are the recent cases of the year 2000 effect and Euro adaptation. In the same line, Rajlich and Bennet (2000) have made a new proposal of a software life cycle oriented towards increasing the importance of SM. These authors consider that, from a business point of view, a software product passes through the following five distinct stages:

- *Initial development*: engineers build the first functioning version of the software product to satisfy initial requirements.
- *Evolution*: engineers extend the capabilities of the software product to meet user needs. Iterative changes, modifications, and deletions to functionality occur.
- *Servicing* (saturation): engineers make minor defect repairs and simple functional changes. During this stage, changes are both difficult and expensive because an appropriate architecture and a skilled work team are lacking.
- *Phase-out* (decline): the company decides not to undertake any more servicing, seeking to generate revenue, or other benefits, from the unchanged software product as long as possible.
- *Closedown*: the company shuts down the product and directs users to a replacement product, if one exists.

Several characteristics change substantially from one stage to another, including staff expertise, software architecture, software decay (the positive feedback, the loss of team expertise, and the loss of architectural coherence) and economic benefits. From the point of view of the SMP, another important difference between one stage and another is the different frequency with which each type of maintenance is carried out. Corrective maintenance (correcting errors) is more usual in the servicing stage, while perfective maintenance (making changes to functionality) is more frequent in the evolution stage. The other two types of maintenance, as defined in the ISO 14764 (1998b) Standard—adaptive (changing the environment) and preventive (making changes to improve the quality properties and to avoid future problems)—are usually considerably less frequent.

Whilst the initial development stage is well documented using numerous recognized methods, techniques and tools, the other four stages (which correspond to the SM) have been studied and analysed to a lesser degree. The attention paid to the development of tools and environments, adapted to the special characteristics of the SMP, has been significantly low.

Human organizations operate based on the business process paradigm—"a collection of interrelated work tasks, initiated in response to an event, that achieves a specific result for the customer of the process" (Sharp & McDermott, 2001). The use of this paradigm is required to approach the SMP in all its scope and complexity, merely extending the technological aspects with management aspects. There are already some proposals in this sense; for example, Cockburn (2000) included the following aspects as minimum requirements to consider in the software development process:

- The people, with certain skills, carry out certain roles in the project, working together in different teams (groups of people).
- These people use methodologies to construct products that conform to certain standards (norms) and satisfy quality measurements (criteria). The processes must also satisfy quality criteria.
- The methodologies require certain skills and tools. The tools facilitate conforming with the standards.
- The teams participate in activities (included in methodologies) that belong to processes encompassed by the project. Each activity undertaken helps to reach a milestone that indicates the progress of the project.

This author calls his proposal a "Big-M Methodology," since it integrates the concepts of methodology in its usual sense (a set of related methods and techniques) and of SEE, as it is defined in the ISO 15940 (ISO/IEC, 2000) Standard; this is, as a collection of software tools used to support software engineering activities. Cockburn's idea is also directly applicable to the SMP, for example, for definite the goals of an integrated environment that helps to manage SM projects with a broader perspective of business processes (Ruiz, Piattini, & Polo, 2001a).

PROCESS-SENSITIVE SOFTWARE ENGINEERING ENVIRONMENTS

Manufacturing processes are special cases of business processes. Although SM is not a typical manufacturing process (due to peculiarities, as such the creative human participation), is does have in common the double facet "production vs management". Therefore, it can be useful to highlight the similarities – more than differences- to understand the activities involved in SM from a more global perspective. The same as with a manufacturing project, a SM project consists of two main, interrelated processes: the production process and the management process. In our case, the production process is the same software engineering process that we know as SM, whereas the management process provides the needed resources for the production process, and controls it. This is possible if the SMP returns information about its behaviour to the management process. There are important relationships between both processes and the exterior environment or project environment; the need for performing SM comes from the exterior world, that is, the project environment is what justifies the existence of the SMP. Moreover, the management must fulfil established rules and standards; the project environment also has an indirect influence on the SMP through the management process. Finally, both processes exploit the corresponding technologies that are provided by the project environment.

This traditional approach can change and improve thanks to the use of the *Software Process Technology*, whose main objective is to control the inherent complexity of a software process through a deep understanding of the self process and through an automatic support by means of a comprehensive work environment, named *Process-sensitive Software Engineering Environments* (PSEE). A PSEE allows the integration of the production (SM in our case) and the management technologies. To achieve it, a PSEE implements, controls, and enhances both the feedback and the feed-forward paths by which the management process controls the production process (Derniame, Kaba, & Warboys, 1999).

In a SM project, a PSEE can provide integrated support to the whole project, that is, to the management and SM processes. The PSEE takes advantage of the software process technology and integrates the management and SM technologies for this purpose. (See Figure 1.)

As can be understood from its name, a PSEE is a special class of SEE, that is, "a collection of software tools that are used to support software engineering activities" (ISO/IEC, 2000). The functionalities of a SEE are described in terms of *services* provided by components of the SEE. In a PSEE these services will be

Figure 1. The impact of PSEE.

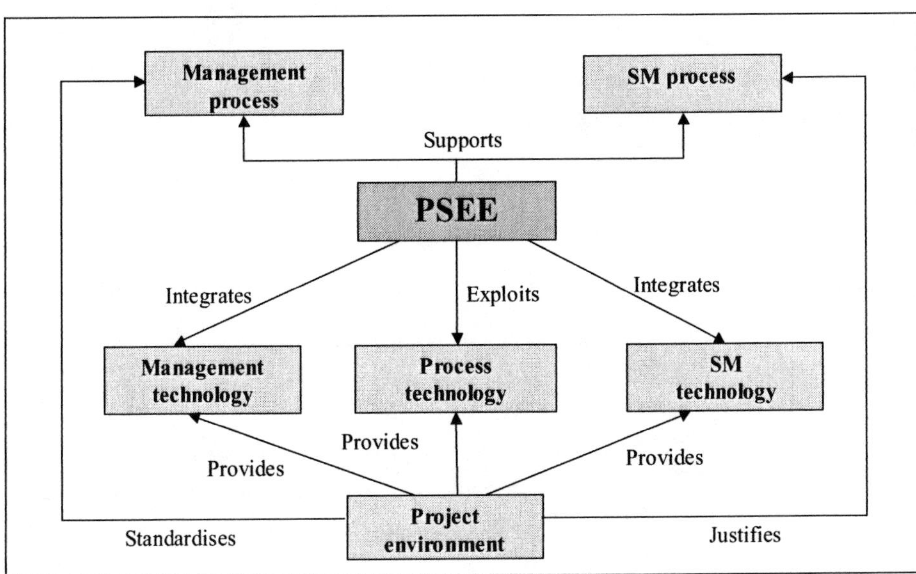

related to the software processes. A service is an abstract description of work done by one or more software tools. A service is self contained, coherent, discrete, and may be a CASE tool. Next, a reference model for architectures in PSEE is presented. The components of this proposal can be derived from the requirements for basic services in a PSEE.

As can been seen in Figure 1, a key feature of a PSEE is the computerized process support, that consists of:

- a process model in a computer-governable format, and
- the necessary tools to define, modify, analyse and enact it.

The essential components of this support are summarized in Figure 2. A repository is used to store the process model together with the product definition data and information on the status of process enactment. Besides the repository, there is another memory level in a PSEE: every actor playing a role in an activity has a workspace (the set of computing resources used by the aforementioned author). A PSEE is controlled by a *process engine* (a process modelling language interpreter). The main goal of this engine is to *control the information flow* among the actors that carry out the activities, according to set up activities in the process model. This control is represented, in the figure, by the broken arrow from the process engine to the communication layer. This information flow takes place between the repository and the workspaces, from some workspaces to others, and among the users and their workspaces. Furthermore, a PSEE must be capable of exchanging information with the environment, for which it must have of a set of import/export channels to exchange data and metadata in an adequate format.

Figure 2. Functional architecture of a PSEE.

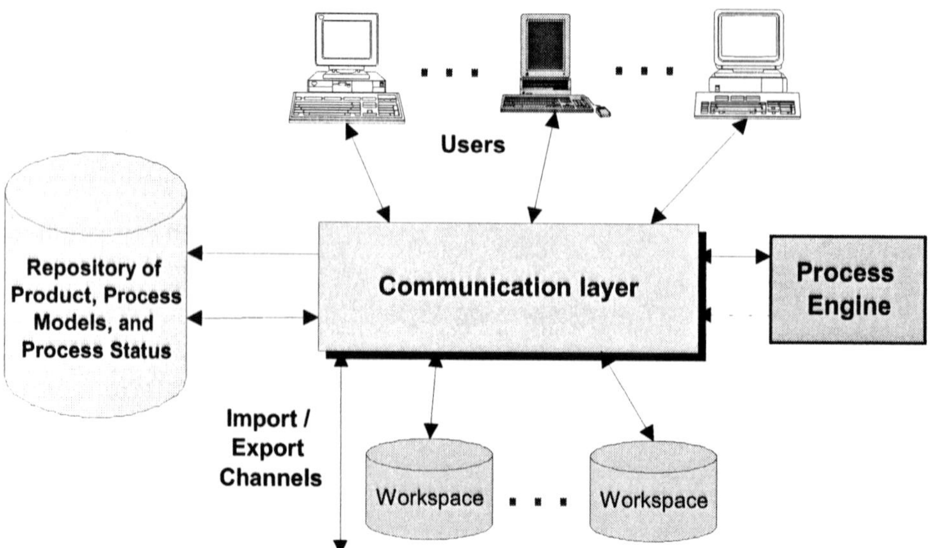

There is an evident similarity between this architecture and that of a computer, because a PSEE is an abstract machine for software development and maintenance. The role of the stored programs in a computer is that played by the enactment models in a PSEE; the equivalent to the main memory is the repository; workspaces in the PSEE are the machine registers in a computer; and tools play the same role of the primitive operations in a computer.

MULTILEVEL CONCEPTUAL ARCHITECTURE

An important principle of modern software engineering is the partitioning of a system in encapsulation layers which can mostly be specified, designed and constructed independently. Following this philosophy, it is very useful to use a multilevel conceptual architecture to manage all the data and metadata of a PSEE.

To manage the complexity of the SMP, and in order to build open, extensible environments, the use of metamodels (models of models) is required. The main uses of metamodels are:

- As conceptual schemas for repositories providing support to the data related to software engineering activities;
- As conceptual schemas for CASE tools;
- To define modelling languages of software processes such as maintenance;
- To allow interoperability between tools (based on interchange standards); and
- As tools for understanding the relationships between concepts and different modelling languages.

There are different proposals related to this type of conceptual architectures for managing models and metamodels: Open Information Model (OIM) of the Meta Data Coalition (MDC, 1999); CASE Data Interchange Format (CDIF) proposed by ISO (ISO/IEC, 2001); and Meta-Object Facility (MOF) proposed by the Object Management Group (OMG, 2000a). In the following, we present MOF, a standard for object-oriented metamodelling, which is useful for defining, representing, and managing metamodels. MOF and its associated specifications include:

- A four-level architecture for metadata that provides a generic pattern for building metadata-centred systems;
- The MOF Model, a standard abstract language for metamodelling that defines different metamodel constructors;
- The MOF metamodels repository that provides a standard service to create, modify, validate and access metamodels;
- The XMI specification that defines an XML-based format for metadata interchange.

The MOF standard has an important aspect to be considered—the concept of *model*. This term is generally used to describe a piece of the real world; that is, the model is an abstraction of the reality that depends on the modeller's point of view. However, the MOF term *model* has a wider meaning, comprising any collection of metadata that fulfills the following conditions:

a) In some manner, it describes interrelated information.
b) It must correspond with a set of rules that define its structure and consistence (based on an abstract language).
c) It has a meaning in a common semantic framework (for example, an ontology).

Therefore, a MOF model is not a model in the usual sense of the word. It does not have to necessarily describe something of the real world nor define things of the modeller's interest.

Metadata themselves are classes of information and can also be described by means of other metadata (that would belong to a higher abstraction level). In MOF terminology, those metadata that describe other metadata are called meta-metadata; then, a model composed of meta-metadata is called *metamodel*. A typical system can be composed by different kinds of metadata (to represent system data, the processes carried out, different user views, etc.). Taking these aspects into account, the framework established by MOF needs to give support to different metamodels. MOF integrates all these metamodels through the definition of a common, abstract syntax. This abstract syntax is known as MOF Model and, therefore, constitutes the common model for all metamodels.

The framework established by the MOF standard is typically described as a four-layer architecture, like that shown in Figure 3. These layers are described as follows:

- The *user-object layer* is comprised of the information that we wish to describe, that is, the *data*.
- The *model layer* is comprised of the metadata that describes information. The metadata are aggregated as *models*.
- The *metamodel layer* is comprised of the descriptions that define the structure and semantics of metadata. This meta-metadata are aggregated as *metamodels*. A meta-model can be seen as a "language for describing different kinds of data."
- The *meta-metamodel layer* is comprised of the description of the structure and semantics of meta-metadata, i.e., it is the abstract language for defining different kinds of metadata.

The main aim of having four layers is to support multiple models and metamodels. As the figure shows, the M1 layer corresponds with the specific models for any application domain. The M2 layer allows the simultaneous work with different types of models (for example, software process, UML, entity/

relationship, relational, etc.). In the latter, the abstract concepts used in the M1 layer are defined: entity, table, attribute, activity, role, etc. Similarly, the M3 layer can describe many other metamodels that in turn represent other kinds of metadata. Summarizing, this four layer metadata architecture has the following principal advantages:

- It can support practically all kinds of imaginable meta-information.
- It allows different kinds of metadata to be related.
- It allows the interchange of both metadata (models) and meta-metadata (metamodels).

The M3 layer is composed of the MOF Model that constitutes the meta-language for metamodel definition. The main modelling concepts provided by MOF are similar to the corresponding ones in UML, although with slight differences:

- *Classes*, which model MOF meta-objects. Classes can have *attributes* and *operations*.
- *Associations*, which model binary relationships between meta-objects.
- *Data Types*, which model other data (e.g., primitive types, external types, etc.).

Figure 3. MOF four layer meta-data architecture.

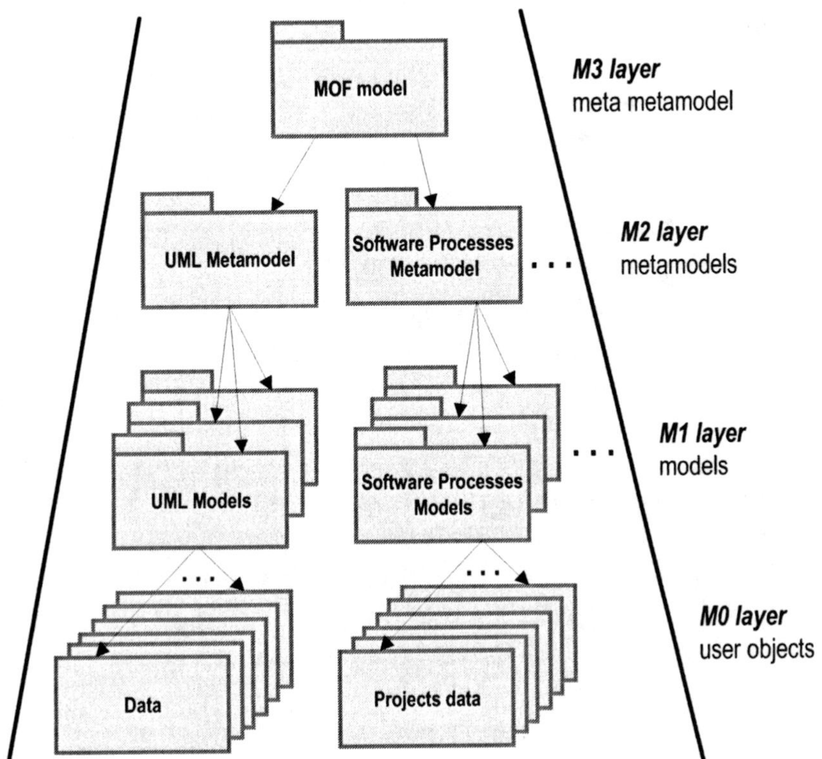

- *Packages*, which modularise the models, are collections of related classes and associations. Their main utility is to facilitate the reuse.

The MOF Model has two important features that distinguish it from other proposals:

- It is object-oriented, supporting metamodeling constructs (class, association, etc.) that are similar to UML's object modelling constructs.
- It is self-describing; it is formally defined using its own metamodeling constructs.

MANTIS BIG-E ENVIRONMENT

Many studies have demonstrated that the majority of the overall expenses incurred by a software product throughout its lifecycle occur during the maintenance stages (Pigoski, 1996), and that the characteristics specific to this stage which differentiate it clearly from the development stage make it very useful to have specific methods, techniques, and tools at one's disposal. Following this line, the MANTIS project aims to define and construct an integrated Big-E Environment for the management of SM projects. By using the nomenclature Big-E Environment, our intention is to emphasize the idea put forth by Cockburn, already mentioned in another section, and integrate and widen the concepts of *Methodology* and *Software Engineering Environment* (SEE). The Alarcos Research Group at the University of Castilla-La Mancha is undertaking the MANTIS project, in collaboration with the company Atos ODS Origin. MANTIS has been partially supported by the European Union, the Inter-ministerial Commission of Science and Technology (CICYT), and the Science and Technology Ministry of Spain (1FD97-1608TIC, FIT-070000-2000-307).

General Characteristics

Probably, the principal advantage of MANTIS is that it integrates practically all the aspects that must be taken into account for directing, controlling, and managing SM projects under one conceptual framework. This advantage is built upon the following features:

- a MOF-based conceptual architecture that facilitates working with the significant complexity inherent in the management of SM projects; specifically, the M3 layer allows work with the different metamodels needed in MANTIS: of processes, of product, of organization, etc.;
- the integration of methods and techniques that have been specially developed for the SMP, such as the MANTEMA methodology (Polo, Piattini, Ruiz, & Calero, 1999a), or adapted from the development process;

- the integration of horizontal and vertical tools by means of orientation to processes, based on the PSEE architecture previously explained, and the use of standards for storage and interchange of data and metadata (specially XMI); and
- a software process metamodel based on the corresponding international standards of ISO, extended with other interesting contributions.

Other requirements that make MANTIS really useful for managing SM projects are:

- *Tool Oriented*: All the MANTIS services are supplied through software tools according to the philosophy of the integrated SEE. The integrated use of all possible tools is advantageous as it increases productivity, reduces the possibility of errors, and facilitates management, supervision and control.
- *Process Driven*: The orientation to processes is a very important vehicle for integration (Randall & Ett, 1995). In MANTIS, two types of integration can be dealt with simultaneously; that of the processes of the organization with the Big-E Environment and that of the tools and artifacts with the processes. Both types of integration are possible thanks to the PSEE architecture.
- *Scalability*: Adaptation to the necessities of projects of any size is also very convenient. The tailoring process of ISO 12207 is recommended for this goal (Polo et al., 1999b).
- *Integrated Support to Technical and Managerial Works*: that is, to the software engineering activities that carry out the SM and to the organizational and managerial activities respectively. This support is necessary to make better use of the management technology, the process technology, and the SM technology (see Figure 1).
- *Role-based*: the organizational metamodel is based on the concept of role. Three roles (maintainer, customer and user) and several sub-roles have been defined (Polo et al., 1999c). Distinguishing between roles (already defined) and responsibilities (that consist of assigning, in each project, tasks to concrete persons and teams) allows the adaptability to the specific characteristics of each organization.

Components

All the MANTIS components are categorized as tools of three different types: conceptual, methodological, and technical (i.e., CASE software tools). A summary of the components that currently make up the MANTIS Big-E Environment is shown in Figure 4. The principal components are described in the following paragraphs, highlighting the most original aspects and those of most interest to organizations that must undertake SM projects.

Figure 4. MANTIS Big-E Environment components.

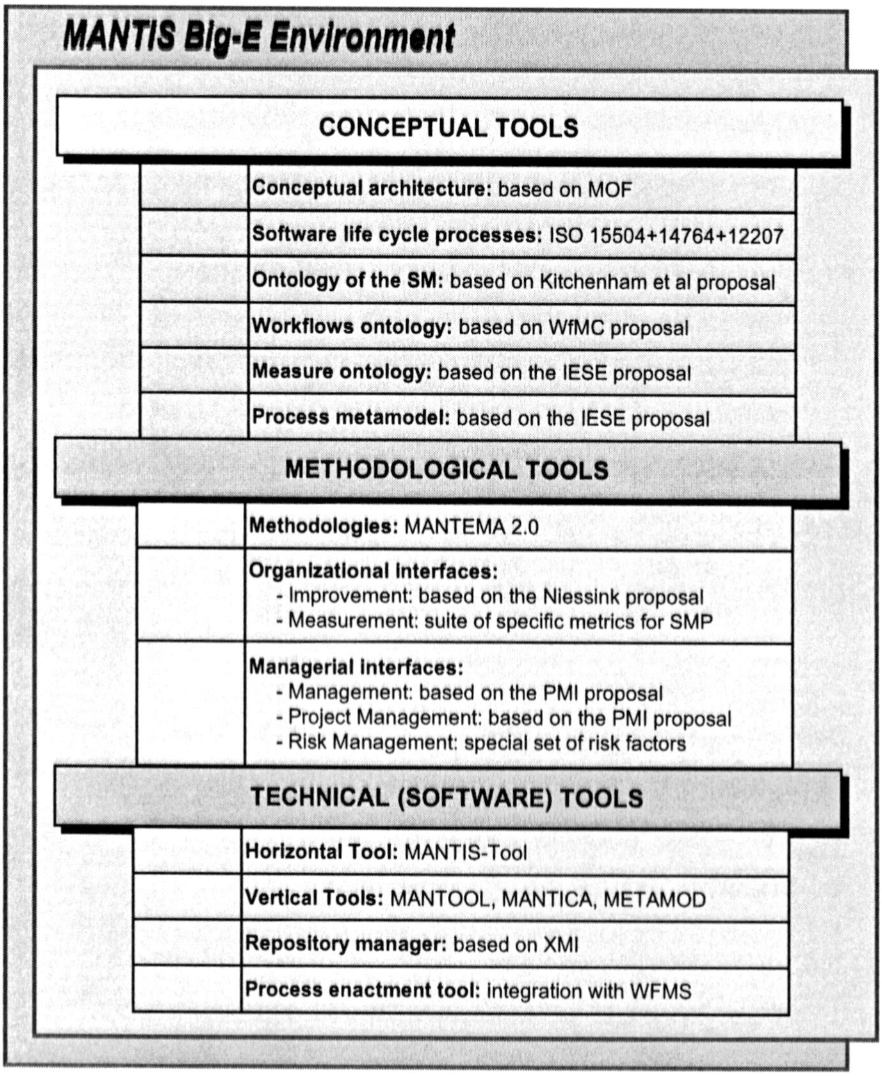

MANTIS Big-E Environment

CONCEPTUAL TOOLS

Conceptual architecture: based on MOF

Software life cycle processes: ISO 15504+14764+12207

Ontology of the SM: based on Kitchenham et al proposal

Workflows ontology: based on WfMC proposal

Measure ontology: based on the IESE proposal

Process metamodel: based on the IESE proposal

METHODOLOGICAL TOOLS

Methodologies: MANTEMA 2.0

Organizational interfaces:
- Improvement: based on the Niessink proposal
- Measurement: suite of specific metrics for SMP

Managerial interfaces:
- Management: based on the PMI proposal
- Project Management: based on the PMI proposal
- Risk Management: special set of risk factors

TECHNICAL (SOFTWARE) TOOLS

Horizontal Tool: MANTIS-Tool

Vertical Tools: MANTOOL, MANTICA, METAMOD

Repository manager: based on XMI

Process enactment tool: integration with WFMS

CONCEPTUAL TOOLS

Conceptual tools are used to represent the inherent complexity of SM projects. A level-based conceptual architecture is necessary to be able to work at different detail levels. A software life cycle process framework is useful for knowing which are the software processes related to the maintenance process. To make sure that all the concepts are correctly defined, used, and represented, a generic ontology for the SM is used. Moreover, in the MANTIS Big-E Environment, two different but complementary points of view of SMP are considered:

- a real-world SMP, that includes all the real activities needed to carry out the maintenance project; and
- an SMP metamodel, which is a representation of the real-world activities for steering, enforcing or automating parts of that SMP.

A real-world SMP is defined with a Workflow ontology based on the workflow technology. A Measure ontology, for SM projects management, has been defined to estimate and improve this process, measuring what is happening with those real-world projects.

In conclusion, an adequate software process generic metamodel is required to represent the different ontologies.

Conceptual Architecture

Four conceptual levels that are based on MOF have been defined. These four levels of the MOF architecture and their adaptation to MANTIS (Ruiz, Piattini, & Polo, 2001b) can be seen in Table 1.

Level M0 has the data of real and specific SM projects with concrete time and cost restrictions. The data handled at this level are instances of the concepts defined at the higher M1 level. The most important specific models that are used at level M1 are based on the MANTEMA methodology and a group of techniques adapted to the special characteristics of the SM. Level M2 corresponds to the SMP metamodel, which will be discussed later in more detail. For example, the generic concept of Activity used in M2 is instanced in the Modification Request Analysis concept in M1 and these, in turn, appear in level M0 as Analysis of the modification

Table 1. Conceptual levels in MANTIS.

LEVEL	*MOF*	*MANTIS*	EXAMPLES	
M3	MOF Model (meta-metamodel)	MOF Model	MOF-class	MOF-association
M2	Metamodel	SMP metamodel & other metamodels	Activity	Artefact is output of "Activity
M1	Model	MANTEMA & other techniques (SMP concrete models)	Modification Request Analysis	Implementation Alternatives Report" is output of "Modification Request Analysis"
M0	Data	Data of SM projects (data of real-world SM projects enactment)	Analysis of the modification request nº 36 of the PATON project	(implementation alternatives) report nº 23 is output of .... in the PATON project

request nº 36 of the PATON project. In the upper conceptual M3 level of MANTIS, the SMP metamodel (and others) are represented in a MOF model. Consequently, all the concepts represented in level M2 are now considered instances of MOF-class or MOF-association. For example, Activity, Actor, or Artifact will be instances of MOF-class; and Activity use Resource or Artifact is input of Activity are instances of MOF-association.

Software Life Cycle Processes

In MANTIS, besides the specific SMP, we need to take into account other sets of processes that support the maintenance management. The software processes framework proposed in the ISO 15504-2 Standard (ISO/IEC, 1998a) has been used to define them. Figure 6 shows the managerial and organizational processes categories; we have underlined those processes that are more important, according to the MANTIS project goals.

Figure 5. Organizational processes for managing the software maintenance.

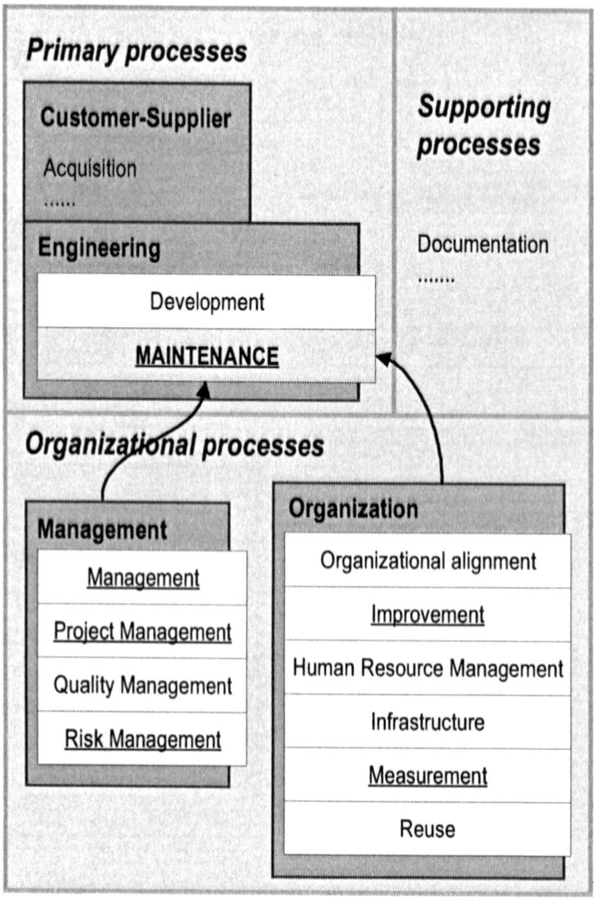

Primary processes
Customer-Supplier
Acquisition
......
Engineering
Development
MAINTENANCE
Supporting processes
Documentation
....
Organizational processes
Management
Management
Project Management
Quality Management
Risk Management
Organization
Organizational alignment
Improvement
Human Resource Management
Infrastructure
Measurement
Reuse

These processes are classified in two categories:

1) *Management category*: consists of processes that contain practices of a generic nature that may be used by anyone who manages any type of project or process within a software life cycle. The purposes of the processes belonging to the Management category—with high importance for MANTIS—are:
   - *Management*: to organize, monitor, and control the initiation and performance of any processes or functions within the organization so they may achieve their goals and the business goals of the organization in an effective manner.
   - *Project management*: to identify, establish, coordinate and monitor activities, tasks and resources necessary for a project to produce a product and/or service meeting the requirements.
   - *Risk management*: to identify and mitigate the project risks continuously throughout the life cycle of a project.
2) *Organization category*: consists of processes that establish the organization's business goals and develops process, product, and resource assets which, when used by the projects in the organization, help the organization achieve its business goals. The purposes of the processes belonging to this category—with high importance for MANTIS—are:
   - *Improvement*: to establish, assess, measure, control and improve a software life cycle process.
   - *Measurement*: to collect and analyse data relating to the products developed and processes implemented within the organizational unit, to support effective management of the processes, and to objectively demonstrate the products' quality.

All these processes compose a process system (according to the definition given by Wang & King, 2000); that is, "an entire set of structured software processes." In order to define the specific SMP the ISO 14764 (ISO/IEC, 1998b) proposal has been used, and has been incorporated into the MANTEMA methodology (explained later). The software life cycle definition proposed in the Standard ISO 12207 (ISO/IEC, 1995) has also been used.

Ontology of the Software Maintenance

In order to achieve a really useful SEE, the integration of all its tools from three dimensions, adding the three traditional ones related to knowledge integration (data, control, and user interface), is required. Truly integrated environments (Falbo, Menezes, & Rocha, 1998) are only possible with an adequate consideration of this fourth dimension. Different abstraction levels (using the previously mentioned conceptual architecture) are not enough to satisfy this requirement; all the models and metamodels that have been used in the problem domain (SMP management) based on the same conceptualisation (set of objects, concepts, entities, and relationships among them, assuming that they exist in the domain) are also required. Moreover, an explicit specification of such conceptualisation is also required; that is, building an ontology (Gruber, 1995).

The elaboration of a common ontology to all the components of the Big-E Environment is a secondary goal of MANTIS for the above mentioned reasons. For this purpose, the informal ontology proposed by Kitchenham et al. (1999) is adequate. A certain formalization level is required in order to represent the ontology by means of objects of the aforementioned conceptual levels, as well as to build the tools that keep and manage models and metamodels. MANTIS uses UML to formalize ontologies. The proposal of Kitchenham et al. is useful for all those

Figure 6. Summarized view of the informal ontology for the SM (Kitchenham et al., 1999).

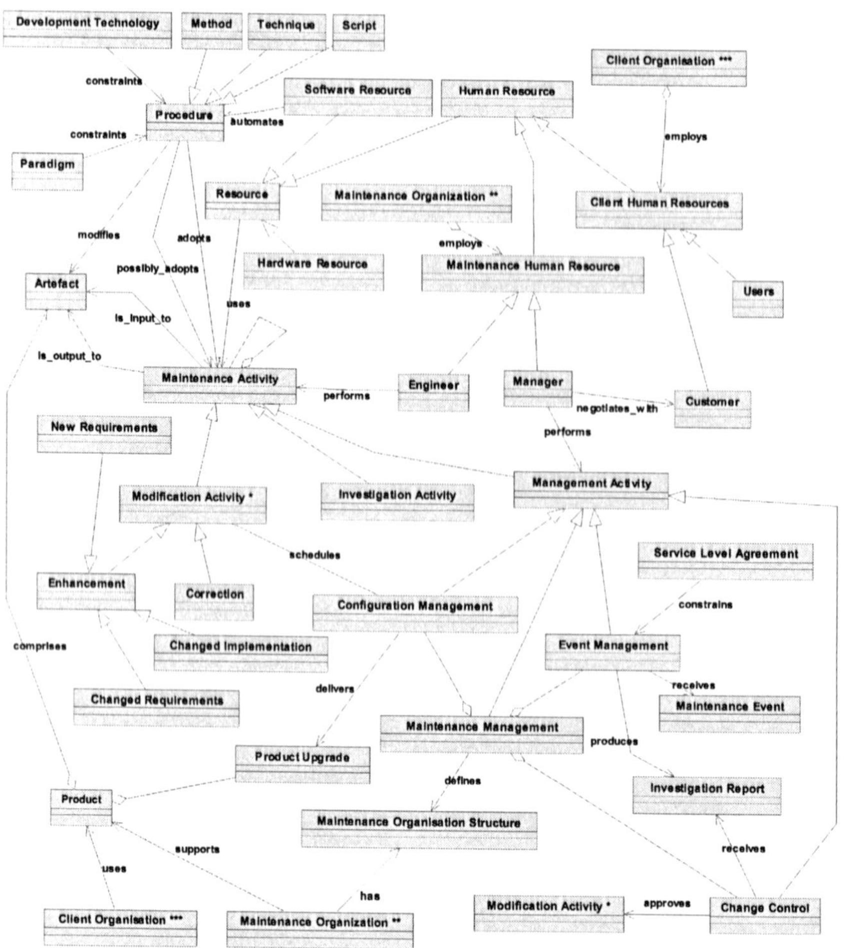

persons that are working in the field of SM and, of course, also for defining and building maintenance-oriented SEE. This proposal is structured in several partial subontologies focused on the *Activities*, the *Products*, the *Peopleware* and the *Processes*. In the MOF-based conceptual architecture used in MANTIS each of these ontologies can be represented with partial metamodels of M2 level that operates like a MOF-package in level M3, allowing the reuse of these metamodels.

Figure 6 shows a summarized and integrated view of these partial ontologies (in UML diagram class format). Due to the size of this proposal, we will only present the partial ontologies that compose it, recommending to the readers refer to the reference (Kitchenham et al, 1999) for more information. In short, each of these ontologies represents the following aspects of SM:

- *Products ontology*: how the software product is maintained and how it evolves with time.
- *Activities ontology*: how to organize activities for maintaining software and what kinds of activities they may be.
- *Processes ontology*: approaches the SMP from two different perspectives, defining a sub-ontology for each one:
  - *Procedures sub-ontology*: how the methods, techniques, and tools (either specific or shared with the development process) can be applied to the activities, and how the resources are used in order to carry out these activities.
  - *Process Organization sub-ontology*: how the support and orgaizational processes (of ISO 12207) are related to the SMP activities, how the maintainer is organized, and what its contractual obligations are.
- *Peopleware ontology*: what skills and roles are necessary in order to carry out the activities, what the responsibilities of each one are and how the organizations that intervene in the process (maintainer, customer and user) relate to each other.

Workflows Ontology

Recently, some authors (Ocampo & Botella, 1998) have suggested the possibility of using workflows for dealing with software processes, taking advantage of the existing similarity between the two technologies. The value of *Workflow Management Systems* (WFMS) in the automation of business processes has been clearly demonstrated and, given that SMP can be considered as part of a wider business process, it is reasonable to consider that workflow technology will be able to contribute a broader perspective to SMP (which we could call *Process Technology*) in line with the objectives of our MANTIS Big-E Environment.

These reasons have led us to integrate the workflow technology in the generic MANTIS ontology. We have incorporated aspects of the Workflow Reference

Model of the Workflow Management Coalition (WFMC, 1995) and aspects of other proposals of workflow metamodels (Liu, Lin, Zhou, & Orlowska, 1999) in level M2 of the conceptual architecture.

We have extended the central part of the SMP generic ontology that represents the existing activities and their performers (see Figure 6) by adding a new partial ontology, called the *Workflows ontology*, which incorporates aspects corresponding to the two following issues:

1) Specification of the structure of activities and sub-activities and their relations (left part of Figure 7); and
2) Information for support, administration, and control of the process enactment (right part of Figure 7).

Activities Specification

Activities specification includes how the activities can broken down into

Figure 7. Summary of the Workflows ontology.

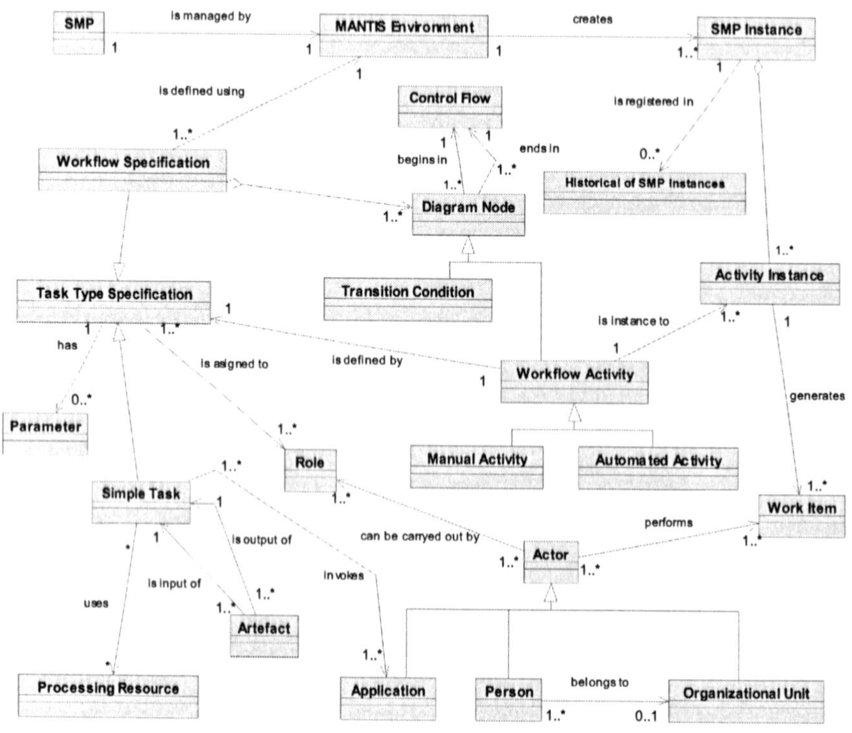

simpler activities, their relationships and control flow, and the possibility of automatic execution. The different roles that can be carried out by each activity are also included. The following objects and relationships are incorporated:

- The SMP itself, is seen as a business process, managed using the MANTIS Environment.
- The Task Type Specification class contains the properties that can be abstracted from an activity or task, and that are independent from workflows; i.e., it includes the general properties of a set of tasks of the same type. There is a distinction between each Task Type Specification and the node representing it in the workflow (Workflow Activity).
- A Task Type Specification can be atomic or nested. The first have no internal structure and are represented through the Simple Task class. The latter have an internal structure that is represented using a Workflow Specification. The execution of a nested task implies the execution of the underlying workflow in it.
- Recursivity is used to represent the task hierarchy. A maintenance project is modelled as a single main Task Type Specification. This main task has an associated Workflow Specification, which includes different Workflow Activities (appearing as Diagram Nodes), each one being defined by a Task Type Specification. In its turn, every one of these specifications has an associated Workflow Specification, which includes other Workflow Activities, and so on. The number of recursivity levels is given by the levels used in the decomposition structure of project. Recursivity finishes when a Task Type Specification corresponds to a Simple Task.
- Each Workflow Specification is represented through the Diagram Node, Control Flow, Transition Condition and Workflow Activity classes. The workflow model used is based on the proposal of Sadiq and Orlowska (1999).
- A Workflow Specification is represented using workflow diagrams (see Figure 8), this is, a set of Diagram Nodes interconnected by means of Control Flows (arrows). Diagram Nodes can be Workflow Activities or Transition Conditions. A condition can be either Or-split or Or-join. Conditions Or-split and Or-join respectively allows to represent branches (choice) and fusions (merge), that is, optional execution paths. To represent concurrent execution paths, activities with more than a control flow beginning in them (beginning of concurrency) or more than a control flow finishing in them (end of concurrency or synchronization) are used.
- The Workflow Activity class has specializations of Manual Activity and Automated Activity, in order to check that automated activities can be carried

out. The difference is in the run-time characteristics. A Workflow Activity can simultaneously be specialized from both subclasses. This possibility allows the existence of mixed activities, that are developed both manual and automatically.

- The Actors involved in the process can be Persons or Organizational Units, to whom the persons belong. Each Actor can play a set of predefined "Roles". In order to take into account the possibility that certain tasks are automatically or semiautomatically performed by invoking external applications, the Application specialization of Actor is included. These Applications can be invoked from Simple Tasks.
- The "Parameter" class allows each Task Type Specification to have some associated parameters. The purpose of these parameters is to manage the information received or produced by the corresponding task type.
- Each Task Type Specification can be assigned to several different Roles. Each Role can be played by several Actors.
- Monitoring and control of a project is made at simple task level. For this reason, the Processing Resources used by a task and the input or output Artifacts are associated to Simple Tasks.

It is interesting to highlight that the inclusion of the Role class helps in the fulfillment of the requirement of that the MANTIS Environment is *role-based*, since a clear distinction between Roles and Actors is established, as well as between the assignment of roles to tasks ("is assigned to" association between Task Type Specification and Role) and responsibilities ("performs" association between Actor and Work Item). With this, a greater flexibility is obtained in order to tailor the environment to organizations and projects with different characteristics and size.

Process Enactment

The dynamic aspects related to the execution are represented through the following objects and relationships:

Figure 8. Diagrams to represent the Workflow specifications.

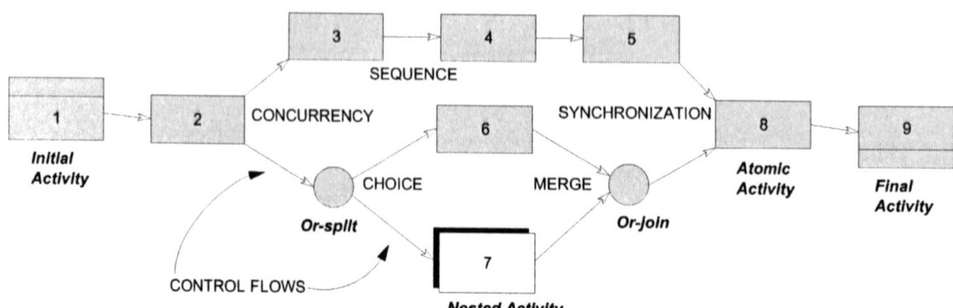

- When a concrete SM project is carried out, the MANTIS Environment creates a SMP Instance, that is registered on a Historical of SMP Instances.
- During the realization of the project, one or more Activity Instances are created for each Workflow Activity. The option to be able to create several run-time instances of a given activity is included to allow iterations.
- Each Activity Instance generates one or more Work Items (the smallest piece of work that is realized, managed, and controlled). Each Work Item is realized by one or more Actors.

Measure Ontology

A fundamental aspect to manage and control any project is the availability of a set of metrics which will allow the measurement of the product that is being produced or maintained and how the project is being executed. Both aspects are fundamental for quality assurance and process assessment or improvement. For these reasons, and for software engineering to be considered as such, it is essential to be able to measure what is being done. For this purpose, the measure ontology of MANTIS includes the concepts of measure, metrics, value and attribute associated to the activities, artifacts, and resources, bearing in mind that the measurements can refer to both product and process aspects. This measure ontology of MANTIS is based on the proposal of the Institut Experimentelles Software Engineering (IESE) that represents a generic schema for the modelling of software processes, in which measurement aspects are very important (Becker, Kornstaedt, & Webby, 1999). The following concepts are taken into consideration:

- The activities, artifacts, resources, and actors are "appraisable elements." In order to be able to measure process enactment, the run-time instances of SMP and activity, and the work items are also appraisable elements.
- An appraisable element has "attributes" that are susceptible to measurement. For example, the duration of an activity or the length of a code module (an artifact type).
- Each attribute has a specific "attribute type," with the possibility that "subtypes" may exist. For example, the "duration of an activity" is a subtype of "quantity of time."
- A "metric" is a formula for measuring certain types of attributes. A "measure" is an association between a specific attribute and a metric. Its principal property is the "value" obtained.

Process Metamodel

In order to be able to represent, in an integrated way, the aforementioned ontologies, a generic software process metamodel, which is represented at level

M2 of the conceptual architecture, is required. In this way, concrete models of processes (M1 level) can be instanced from this metamodel. In MANTIS, the proposal of the IESE (Becker & Webby, 1999) is used for this goal and also for the measure ontology.

The IESE proposal considers that the main aspects to be taken into account when modelling processes are:

- how things are done,
- who does them, and
- what is produced, used, or modified during the activities.

In Figure 9, the basic components which should be included in any process metamodel are shown. As can be seen, within the generic software processes metamodel, three partial metamodels stand out:

1.) Process Modelling,
2.) Measurement, and
3.) Human Resources.

Process Modelling

This metamodel is prepared from the necessary constructors to define models of software processes. For it, the Element abstract class is the root class, from which Entity and Relationship classes are defined by inheritance (in a similar manner to the use of MOF-class and MOF-association in the MOF model).

Figure 9. Generic metamodel for software processes.

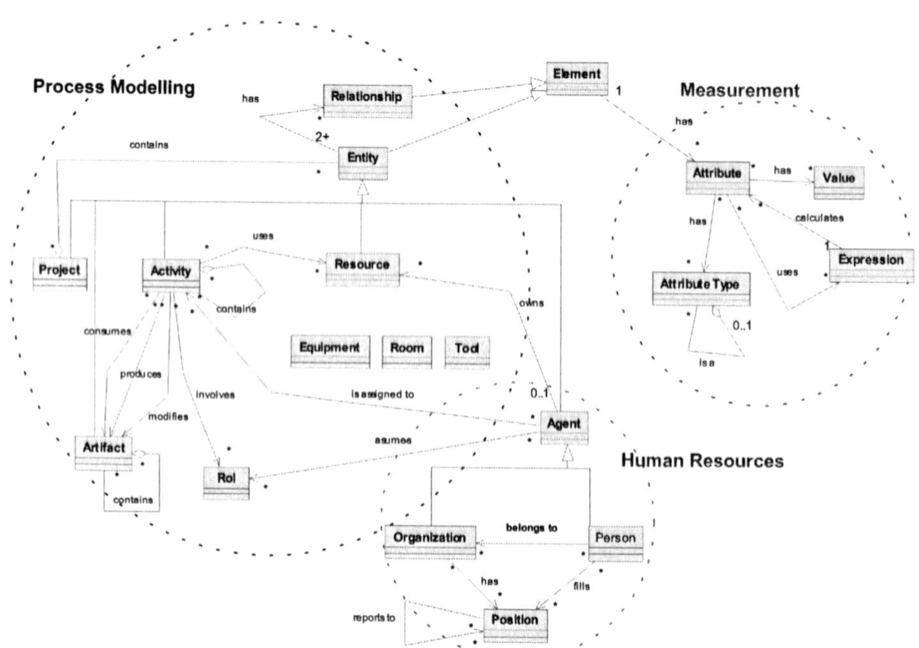

Every model of processes is made up of a set of Activities, which in turn can include other activities. An activity is an abstraction of *how things are done*. An activity is a step in a process that may contain anything from activities of software development or maintenance to project management or quality assessment activities. Entity is a superclass that is specialized in the Project, Activity, Artifact, Resource, and Agent subclasses. The metamodel allows that the same entity can be included in more than one project. Among the other classes which stand out in the metamodel are:

- Agent, which is an abstract class that describes the actors who perform one or more activities;
- Resource, which describe the necessary entities to carrying out the project; and
- Artifact, which is an abstraction of the products that are modified, used, or produced in a project.

Another important characteristic of this partial metamodel is that it is possible to represent the relationships between entities by means of the Relationship class. This class is similar to the MOF-association, but with the difference that it can have a degree greater than two and that it models relationships between entity instances instead of between class instances.

Measurement

In order to represent the aforementioned measure ontology, we must include the possibility of defining indicators for any element in the metamodel. These indicators must allow us to check the quality of the processes, for example, to be able to apply improvement plans. The partial measurement metamodel (mentioned in Figure 9) includes the Attribute, Attribute Type, Value and Expression classes. Each Element of the model can have certain associated Attributes. In turn, each Attribute has an Attribute Type (integer, date, string, ...) and a Value. Examples of attributes are: the duration of an activity, the number of detected errors in a module, and the cost of a resource. In addition, the calculation of an attribute value can be based on an Expression. The same expression can be used with various attributes.

Human Resources

This partial metamodel is outstanding in the generic metamodel since human resources are the key element for successfully performing software processes. With this partial metamodel, it is possible to model who is performing the activities. The following classes stand out:

- Role, which is an abstraction of the abilities and necessary conditions to perform an activity. Each activity is associated with certain roles in the project and Agents assume these roles.
- Organization, which describes the context from the administrative point of view in which the software process occurs. Examples of organization are the software maintenance team, the testing team or a software company.
- Person, which is an individual human agent, for example a programmer or software engineer.
- Position, which describes a post in an organization. A person can hold several positions and a position can be held by several people. Besides, all people have a position in an organization, for instance, the manager in the testing team.

Recently, a new proposal for software process metamodelling has been published. An OMG adopted specification, named *Software Process Engineering Metamodel*, was published in December-2001, being its definite approval scheduled for July-2002. This document presents "a metamodel for defining processes and their components, but the enactment of processes .... is not in the scope of this metamodel." At the core of this proposal is the idea that a software process is a collaboration between abstract active entities called process *roles* that perform operations called *activities* on concrete, tangible entities called *work products* (equivalents to artifacts in MANTIS). The possibility of using this metamodel in MANTIS instead of the IESE proposal is being studied at this time.

METHODOLOGICAL TOOLS

Because maintenance is the most expensive stage in the software life cycle, it is important to have methodologies and tools so that we can approach this problem in the best possible way. Since necessities in the development and maintenance phases are practically divergent and different (see Figure 10), it becomes necessary to grant to the SM the importance that it has, keeping in mind its special characteristics and its differences with the development phase. The MANTEMA methodology was developed with this objective in mind (Polo et al., 1999a). For these reasons, MANTEMA 2.0 is the proposed methodology in the MANTIS Environment. For it, a MANTEMA process model is included in M1 level of the conceptual architecture.

For an adequate management of the SMP, MANTIS considers interfaces with several organizational and managerial processes (see Figure 5): improvement, measurement, management, project management, and risk management. For this issue, the models of these different processes also can be represented at the M1 level of the conceptual architecture.

Figure 10. Development versus maintenance efforts.

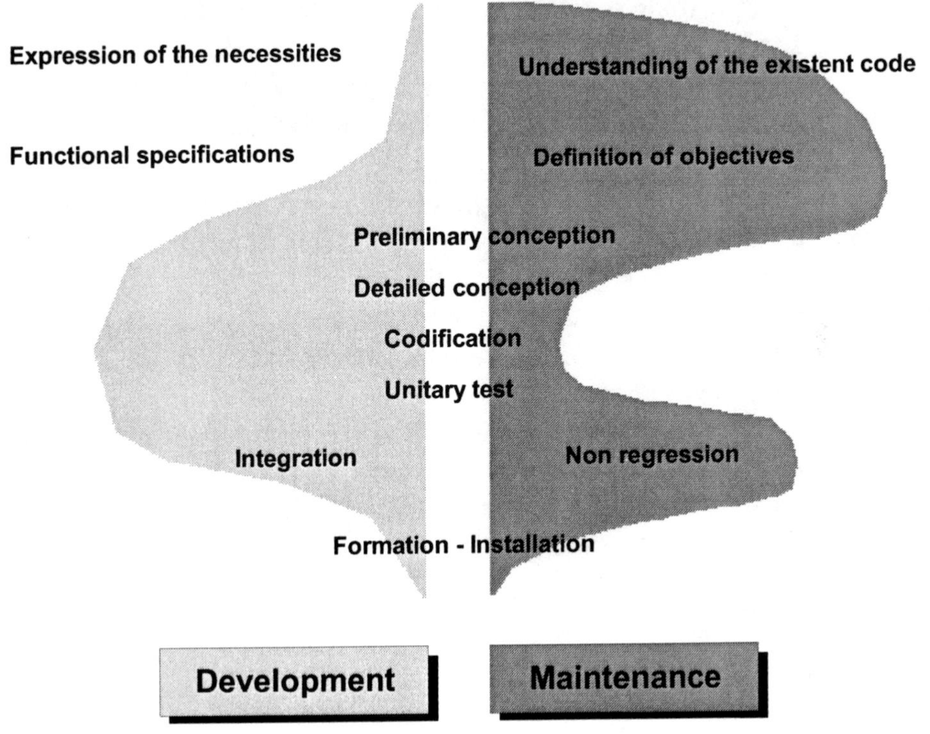

MANTEMA Methodology

The readers have a whole chapter dedicated to MANTEMA in this same book and, therefore, this section only explains the role model used, given its importance in process management issues. MANTEMA classifies the roles depending on the organization to which belong (Polo et al., 1999c):

- *Customer organization*, which owns the software and requires the maintenance service. Its roles can be:
  - The *Petitioner*, who promotes a modification request, establishes the needed requirements for its implementation, and informs to the maintainer.
  - The *System organization* is the department that has a good knowledge of the system that will be maintained.
  - The *Help-Desk*, this is the department that attends users. It also reports the incidents sent by users to generate the modification request to the Petitioner.

- *Maintainer Organization* supplies the maintenance service. Its roles can be:
  - The *Maintenance-Request Manager* decides whether the modification requests are accepted or rejected and what type of maintenance should be applied. He/she gives every modification request to the Scheduler.
  - The *Scheduler* must plan the queue of accepted modification requests.
  - The *Maintenance Team* is the group of people who implement the accepted modification request. They take modification requests from the queue.
  - The *Head of Maintenance* prepares the maintenance stage. He/she also establishes the standards and procedures to be followed with the maintenance methodology used.
- *User Organization* uses the maintained software. Its roles can be:
  - The *User* makes use of the maintained software and communicate to the incidents to the Help-Desk.

This enumeration is not intended to be rigid, since it may be tailored to every particular case including new roles or modifiying the existing ones. Each one of the three organisations listed above may be a different organisation, but this is not always so. Sometimes two or more different roles may coincide in the same actor (person or organizational unit).

Organizational Interfaces

SMP improvement can be managed with MANTIS from the two different perspectives proposed by Niessink (2000):

- *Measurement-based,* the measurement is used as an enabler of improvement activities (see Figure 11); and
- *Maturity-based,* the organization or processes are compared with a reference framework that is assumed to contain the correct activities for the organization or processes. The best-known examples of such reference frameworks are the Software Capability Maturity Model (CMM) and the ISO 15504 Standard.

In MANTIS, we have used the ISO 15504 as a model for the assessment and improvement of the SMP (García, Ruiz, Piattini, & Polo, 2002). For example, Table 2 shows the integration of the ISO 15504 assessment model into the MANTIS conceptual architecture.

The multilevel conceptual architecture of MANTIS enables us to work with different models of one process, which is a requirement in order to be able to manage the process improvement.

On the other hand, the measure process management is also possible with MANTIS thanks to the inclusion of the Measure metamodel previously mentioned,

Figure 11. A model for measurement-based improvement (Niessink & Vliet, 1999).

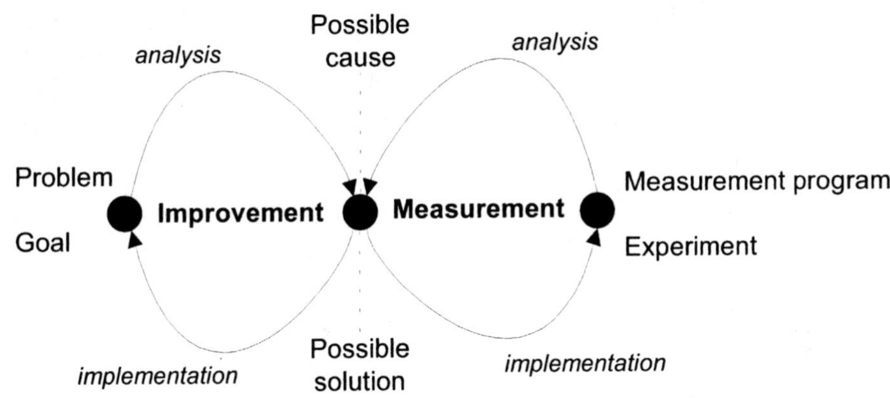

and to the definition of a set of metrics that are suitable for the management of the SMP. In the MANTEMA methodology, metrics are enumerated for each task. Furthermore, some of these metrics are used to define *Service Level Agreements* (SLA) in SM contracts and to verify their later fulfilment:

- Time of Resolution of Critical Anomalies (TRCA): it is the maximum time that the maintenance organization may take in fixing a critical anomaly without being sanctioned.

Table 2. Mapping between generic metamodel and assessment model.

M2 Classes	M2 instances (M1 Classes)	
Activity	DAI	Defining the assessment input
	PAP	Perform the assessment process
	PAP.P	Planning
	PAP.CD	Collect Data
	PAP.VD	Validate Data
	PAP.PR	Process Rating
	PAP.R	Reporting
	RAO	Record the assessment output
Artifact	Process	Process to Assess
	Profile	Process Profile
Activity contains Activity		PAP contains Planning
		PAP contains Collect Data
		PAP contains Validate Data
		PAP contains Process Rating,
		PAP contains Reporting
Activity consumes Artifact		PAP consumes Process to Assess
Activity produces Artifact		PAP produces Process Profile

- Time of Resolution of Non-Critical Anomalies (TRNCA).
- Number of Modification Requests related to Critical Anomalies assumable per a time period (NMRCA).
- Number of Modification Requests related to Non-Critical Anomalies assumable per period (NMRNCA).
- Maximum Deviation in the number of Critical Anomalies (DevCA) in a shorter period, or the maximum number of MR that can be received in, for example, a week, with the commitment of being served in TRCA (e.g., NMRCA can be 20 and DevCA=10; this means that the Maintenance Organization commits to serve 20 MR per month, but with the number of MR per week limited to 10).
- DevNCA, which is similar to DevCA, but relates to Non-Critical Anomalies.

Managerial Interfaces

The management and project management process models are based on the ISO 16326 proposal (ISO/IEC, 1998c). Additionally, as in the aforementioned Workflows ontology, a Work Item is defined as the smallest unit of a job which is undertaken and that is controlled, managed, and assigned to an actor to carry it out. This method for controlling the execution of the job is based on the norm of the Project Management Institute (PMI, 2000).

Knowledge of the main risks associated to SM projects is required in order to manage the risk management process with MANTIS. For this, a tailoring of the proposal Control Objectives for Information and Related Technology (CobiT) which has been published by the Information Systems Audit and Control Foundation (ISACF) (ISACF, 1998) has been made. The high-level control objectives list of CobiT has been modified, substituting the Manage Changes for Manage the Software Maintenance Process, and the resulting list of detailed control objectives has been also restructured in terms of the SMP activities and tasks. The result is the high and detailed control objectives (Ruiz et al., 2000) shown in Table 3.

TECHNICAL TOOLS

Although the development of software tools is not the main goal of the MANTIS project, the definition and construction of some has been, in order to support the automatization of SM projects management. The most significative are mentioned below.

Horizontal Tool

MANTIS Tool is a software tool whose functionality is to offer an integrated user interface for using the different vertical tools, as well as any other used software

Table 3. Control objectives for the risk management of SM projects.

High-Level Control Objective:	**AI06 – Manage the software maintenance process**
Description:	*Business activities are performed without accidental interruptions and the software of the existent information systems is adapted to the new necessities.*
Detailed Control Objectives:	
Changes in the operating environment:	An organised procedure exists in order to carry out the migration of a software product from an old operating environment to another new one.
Software retirement:	The methodologies of software development and maintenance include a formal procedure for the retirement of a software product when it has concluded its useful lifecycle.
Maintenance types:	The software maintenance types are categorised, and the activities and tasks to perform each type have been planned.
Maintenance agreement:	The relationships between maintainer and client, and the obligations of each, are established in a maintenance agreement or contract.
Improvement of the process quality:	The methodology used for the SM includes techniques that increase the maintainability (ease of maintenance).
Planning of maintenance:	A maintenance plan exists. It includes the scope of the maintenance, who will perform it, an estimate of the costs, and an analysis of the necessary resources.
Procedures for modification requests (MRs):	Procedures to begin, to receive, and to register MRs exist.
Managing and monitoring the changes:	The maintainer has established an organisational interface between the maintenance and the configuration management process, so that the second can give support to the first.
Analysis and assessment of the MRs:	The MRs are categorised and prioritised, and there are well-structured mechanisms to evaluate their impact, cost, and criticality.
Verification of the problems:	The maintainer replicates or verifies that the problem, which is the source of the MR, really does exist.
Record of the MRs:	The maintainer documents and records the MRs, with the analyses, assessments, and verifications performed.
Approval:	Depending on the maintenance type of an MR, there are formal procedures that detail the approval fashions that the maintainer should obtain, before and after implementing the modification.
Modifications implementation:	To implement the modifications, the maintainer uses the same previously established methodology for the software development process, now adapted to the maintenance process.
Update the documentation:	The documentation (technical reports, manuals, etc.) affected by an MR is updated after finishing the modification.

(WFMS, etc.). Figure 12 shows a summary of its interaction with other types of software tools used in MANTIS.

Vertical Tools

MANTOOL allows the management of modification requests (MR) from their inception until the end of their respective sets of activities and tasks, according to the five maintenance types and the different stages defined in the MANTEMA methodology. Templates of MR and of other documents generated during the maintenance process are carefully detailed. The pursuit of every MR is done on a screen such as that in Figure 13. In this screen, there is a tab for every task belonging to the maintenance type of the MR: inputs, outputs, and metrics of the task are specified on this screen. There is also an additional tab (the one in foreground in Figure 13) that shows the general information of the MR (application, date and time of presentation, number of modification request, last executed task, the user who presented it, and a description of the error and error messages). There is also a graph with nodes associated with every stage. Nodes change their colour when their respective tasks are executed in order to provide a quick idea of the MR status.

The data saved in MANTOOL can be used to extract different kinds of reports and to do estimates of future maintenance interventions: summaries of the

Figure 12. Operation schema of the MANTIS environment software tools.

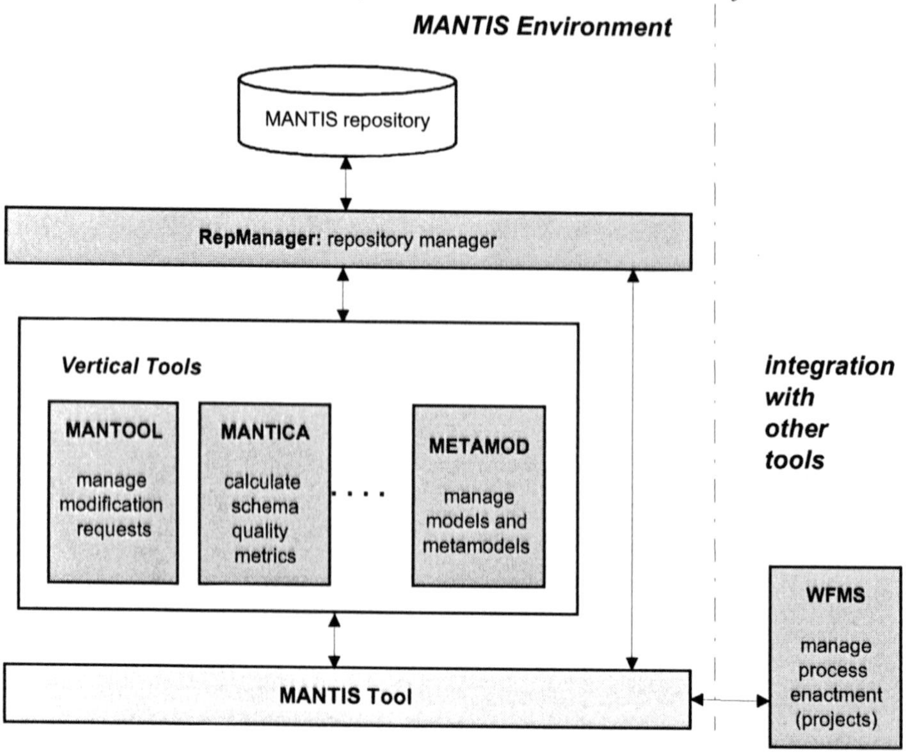

different data saved for every application in every type of maintenance; data of maintenance requests related to a definite type of maintenance; dedication of personnel; deviations (difference between the time initially scheduled for executing an MR and the real time dedicated); tendency (evolution of the different metrics both of modules and of routines); etc.

MANTICA is another tool of the MANTIS environment, developed to define and register quality metrics of relational, object-relational, conceptual, or UML schemas. For other types of metrics, those developed by other authors have been selected.

METAMOD is a MOF-based tool for representing and managing software process models and metamodels (Ruiz et al., 2001c). The application is composed of a metamodel administrator as its principal component and of a graphical user interface that allows a visual description of the classes that make up the core of the MOF model (Package, Class, Datatype, Attribute, Operation, Reference, AssociationEnd, and Constraint). The metamodel administrator has a three-shape structure, as does the MOF model; a package contains classes and associations, a class contains attributes and operations, an association contains restrictions, etc. In Figure 14, the window associated with the MOF-class definition is visible. The

Figure 13. MANTOOL: screen showing the status of a modification request.

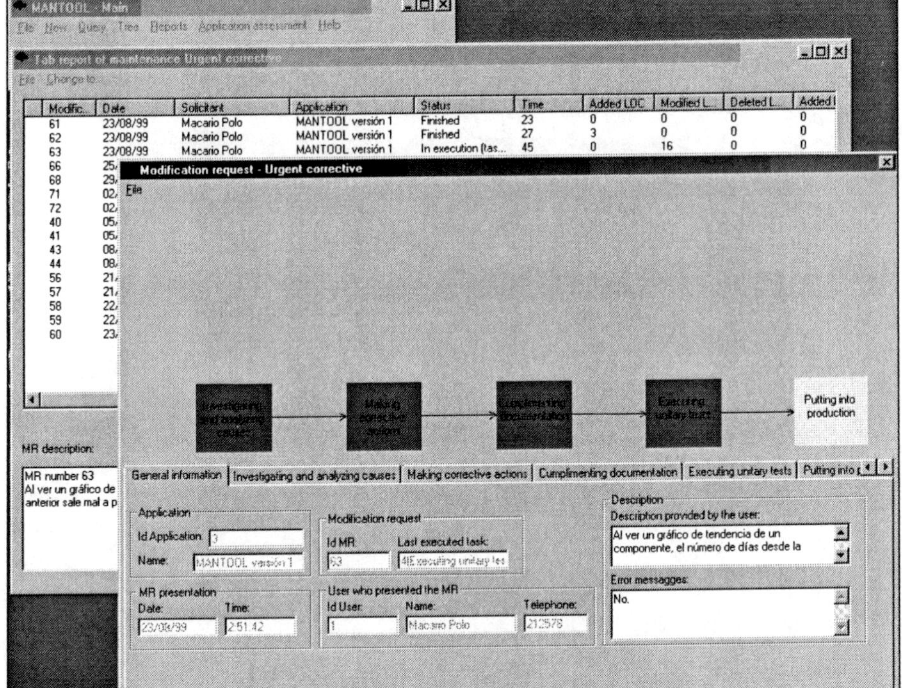

metamodels are validated and internally represented according to the class hierarchy described by the MOF proposal (specified in the IDL attached document).

To complement this tool, other tools of visual modelling based on UML can be used for the visualization of the MOF models as class diagrams (e.g., Rational Rose). This functionality is possible owing to the fact that the tool has the capacity for interchanging MOF models in XMI format (OMG, 2000b).

Repository Manager

A basic aspect for the effectiveness of MANTIS is the existence of a common repository for all the environment components. In order to have an open format for the storage of the data and metadata, the MANTIS repository uses *XML Metadata Interchange* (XMI). It is essential that both the metamodels defined with the MANTIS tools and the metamodels defined with other tools that support XMI are usable. XMI constitutes the integrating element for metadata originating from different sources, as it represents a common specification. Therefore, to a great extent, the RepManager component provides a tool with support enabling it to interchange metadata with other repositories or tools that support XMI. These functions are supported via a set of calls to the system and basically are:

1) Storage of MOF models in the local metadata repository using XMI documents; and
2) Importation/exportation of models and metamodels.

Figure 14. Defining models and metamodels with METAMOD.

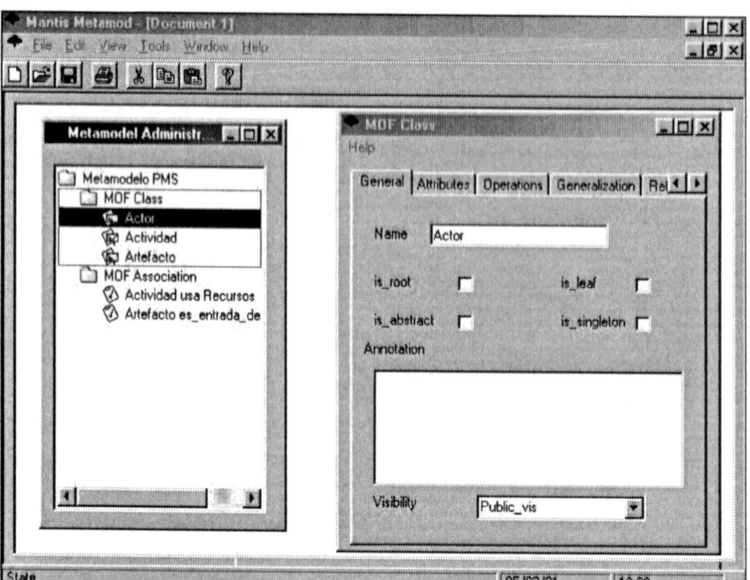

One of the objectives of MANTIS is the integration of the concepts on which it is based and, more importantly, the correct management of all the metadata and data in the different conceptual levels previously mentioned. In order to achieve this, the models and metamodels of the different levels can be stored as XMI documents in the repository managed by RepManager. Each XMI document stored in the repository represents a correspondence between M_i level and M_{i-1} level, as each XMI document will contain the metadata (level M_i) that describe the corresponding data (M_{i-1} level instances). As a result, three types of XMI/XML documents are stored in the repository (see Figure 15):

- Correspondence between the M3-M2 levels, such as the correspondence between the MOF-model and the SMP generic metamodel. The DTD or scheme that represents this type of XMI documents is unique and will be the DTD of the MOF-model.
- Correspondence between the M2-M1 levels, such as the mapping between the SMP generic metamodel with the specific MANTEMA process model. In this case, the DTD represents the metamodel of level M2; for example, it would be necessary to store the DTD of the SMP generic metamodel. Another example of this type is the correspondence between the generic SMP metamodel and the assessment model shown in Table 2.

Figure 15. The three mapping layers of MANTIS represented as XMI/XML documents.

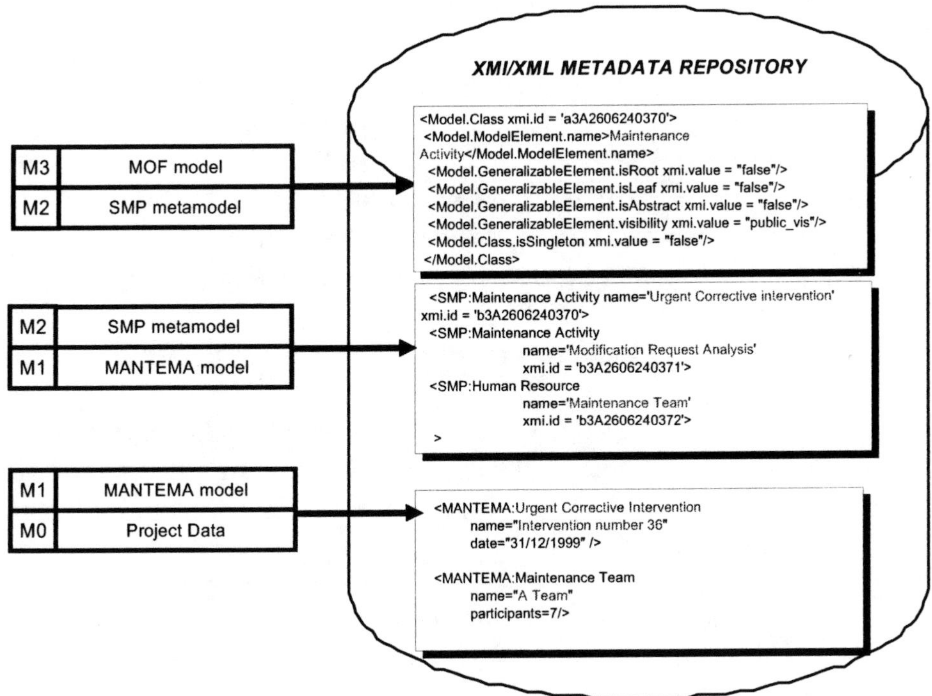

- Correspondence between the M1-M0 levels, such as the mapping between the MANTEMA model and a specific project carried out using that methodology. In this case, the DTD represents specific models of M1 level. In the MANTIS environment case, it would be necessary to store the DTD corresponding to the MANTEMA model.

Process Enactment Tools

Different authors have recommended the use of workflow tools to manage software processes. In (Aversano, Betti, DeLuca, & Stefanucci, 2001), the use of a Workflow Management System for the management of the SMP is analysed. In particular, these authors present a case study that consists of using a WFMS for the management of the modification requests (in a similar way to that provided by MANTOOL in the MANTIS case, but with no distinction of maintenance types). MANTIS suggests the use of a WFMS mainly for the enactment of the processes in each concrete project (see Figure 12). If the WFMS is MOF-based and XMI compatible, interchanging the process models between the WFMS and MANTIS will be possible and the huge potential offered by the workflow technology in the organization, scheduling, monitoring and control of the SM projects will be well utilized.

REFERENCES

Aversano, L., Betti, S., De Lucia, A., & Stefanucci, S. (2001). Introducing workflow management in software maintenance processes. *IEEE International Conference on Software Maintenance* (ICSM), pp. 441-450.

Becker-Kornstaedt, U. & Webby, R. (1999). *Comprehensive Schema Integrating Software Process Modeling and Software Measurement.* Fraunhofer Institute, IESE report Nº 047.99/E., v. 1.2, 1999.

Bennett, K. & Rajlich, V. (2000). Software maintenance and evolution: A roadmap. *International Conference on Software Engineering* (ICSE) - Future of SE Track, pp. 73-87.

Cockburn, A. (2000). Selecting a projects methodology. *IEEE Software*, July/August, pp. 64-71.

Derniame, J.C., Kaba, B.A., & Warboys, B. (1999): The software process: Modelling and technology. In Derniame et al (eds.), *Software Process: Principles, Methodology and Technology*. LNCS 1500, pp. 1-13. New York: Springer-Verlag,

Falbo, R.A., Menezes, C.S., & Rocha, A.R. (1998). Using ontologies to improve knowledge integration in software engineering environments. *Proceedings of the 4th International Conference on Information Systems Analysis and Synthesis*, SCI'98/ISAS'98, Orlando, FL. USA, July.

García, F., Ruiz, F., Piattini, M., & Polo, M. (2002) Conceptual architecture for the assessment and improvement of software maintenance. *Proceedings of the 4th International Conference on Enterprise Information Systems* (ICEIS'02). Ciudad Real (Spain), April.

Gruber, T. (1995): Towards Principles for the Design of Ontologies used for Knowledge Sharing. *International Journal of Human-Computer Studies*, 43(5/6), pp 907-928.

Harrison, W., Ossher, H., & Tarr, P. (2000). Software engineering tools and environments: A roadmap. *International Conference on Software Engineering (ICSE) - Future of SE Track*, pp. 261-277.

ISACF (1998). *CobiT: Governance, Control and Audit for Information and Related Technology*, 2nd edition. Information Systems Audit and Control Foundation. Rolling Meadows, IL.

ISO/IEC (1995). IS 12207 *Information technology - Software life cycle processes*. Geneva, Switzerland: International Organization for Standards.

ISO/IEC (1998a): TR 15504-2 *Information technology - Software process assessment - Part 2: A reference model for processes and process capability*, August.

ISO/IEC (1998b). FDIS 14764 *Software engineering - Software maintenance*, December.

ISO/IEC (1998c). DTR 16326 *Software engineering – Project management*, December.

ISO/IEC (2000). JTC1/SC7/WG4 15940 *Information technology –Software engineering environment Services*, July.

ISO/IEC (2001): FDIS 15474-1 *Software Engineering - CDIF Framework - Part 1: Overview*, working draft 5, March 2001.

Kitchenham, B.A., Travassos, G.H., Mayrhauser, A., Niessink, F., Schneidewind, N.F., Singer, J., Takada, S., Vehvilainen, R., & Yang, H. (1999). Towards an ontology of software maintenance. *Journal of Software Maintenance: Research and Practice*. 11, pp. 365-389.

Liu, C., Lin, X., Zhou, X., & Orlowska, M. (1999). Building a repository for workflow systems, *Proceedings of the 31st International Conference on Technology of Object-Oriented Language and Systems*. IEEE Computer Society Press, 1999, pp. 348-357.

MDC (1999). Meta Data Coalition, *Open Information Model*, v.1.0, August.

Niessink, F. (2000). Perspectives on improving software maintenance. PhD Thesis, Vrije Universiteit, Netherland. Available in http://www.opencontent.org/openpub/.

Niessink, F. & Vliet, H.v. (1999). Measurements should generate value, rather than data. *Proceedings of the Sixth IEEE International Symposium on*

Software Metrics (METRICS'99). Boca Raton, FL. pp. 31-38. New York: IEEE Computer Society Press.

Ocampo, C. & Botella, P. (1998). *Some Reflections on Applying Workflow Technology to Software Processes.* TR-LSI-98-5-R, UPC, Barcelona, Spain.

OMG (2000a). Object Management Group, *Meta Object Facility (MOF) Specification*, v. 1.3 RTF, March. Available in http://www.omg.org.

OMG (2000b): Object Management Group, *XML metadata interchange* (XMI), v. 1.1, November.

OMG (2001). Object Management Group, *Software Process Engineering Metamodel (SPEM) Specification*, December.

Pigoski, T.M. (1996). *Practical software maintenance. Best practices for managing your investment.* New York: John Wiley & Sons.

PMI (2000): *A guide to the project management body of knowledge, 2000 edition.* Newtown Squares, PA: Project Management Institute Communications.

Polo, M., Piattini, M., & Ruiz, F. (2001). MANTOOL: A tool for supporting the software maintenance process. *Journal of Software Maintenance and Evolution: Research and Practice*; Vol 13, Nº 2, pp. 77-95.

Polo, M., Piattini, M., Ruiz, F., & Calero, C. (1999a). MANTEMA: A complete rigourous methodology for supporting maintenance based on the ISO/IEC 12207 Standard. *Third Euromicro Conference on Software Maintenance and Reengineering* (CSMR'99). Amsterdam (Netherland), 1999, pp. 178-181. New York: IEEE Computer Society Press.

Polo, M., Piattini, M., Ruiz, F., & Calero, C. (1999b): Using the ISO/IEC tailoring process for defining a maintenance process. *IEEE Conference on Standardization and Innovation on Information Technology.* Aachen, Germany. pp. 205-210. New York: IEEE Computer Society Press.

Polo, M., Piattini, M., Ruiz, F., & Calero, C. (1999c): Roles in the maintenance process. *ACM Software Engineering Notes*; vol 24, Nº 4, pp. 84-86.

Rajlich, V.T. & Bennett, K.H. (2000): A staged model for the software life cycle. *IEEE Computer*, July, pp. 66-71.

Randall, R. & Ett, W. (1995). Using process to integrate software engineering environments. *Proceedings of the Software Technology Conference*, Salt Lake City, UT. In http://www.asset.com/stars/loral/pubs/stc95/psee95/psee.htm.

Ruiz, F., Piattini, M., Polo. C., & Calero, C. (2000). Audit of software maintenance process. In *Auditing Information Systems*, pp. 67-108. Hershey, PA: Idea Group Publishing.

Ruiz, F., Piattini, M., & Polo, M. (2001a): Using metamodels and workflows in a software maintenance environment. *VII Argentine Congress on Computer Science* (CACIC'01). El Calafate, Argentina, CD of proceedings.

Ruiz, F., Piattini, M., & Polo, M. (2001b). A conceptual architecture proposal for software maintenance. *13th International Symposium on System Integration* (ISSI'01). Baden-Baden, Germany, August, pp. VIII:1-8.

Ruiz, F., García, F., Márquez, L., Piattini, M., & Polo, M. (2001c). Tool based on MOF for software process metamodelling. *Business Information Technology Management* (BitWorld'2001). Cairo, Egypt, CD-ROM.

Sadiq, W. & Orlowska, M.E. (1999). On capturing process requirements of workflow based business information systems. *3rd International Conference on Business Information Systems*, Poznan, Poland, pp. 195-209.

Sharp, A. & McDermott, P. (2001). *Workflow modeling. Tools for process improvement and application development.* Norwood, MA: Artech-House.

Wang, Y. & King, G. (2000). *Software Engineering Processes: Principles and Applications*. Boca Raton, FL: CRC Press.

WFMC (1995). TC00-1003 1.1: Workflow Management Coalition. *The Workflow Reference Model*, January.

About the Authors

Lerina Aversano received the Laurea degree in Computer Engineering from the University of Sannio, Italy, in 2000. She is currently a PhD student at the University of Sannio. Her research interests include process and information system modelling, workflow management, document management, software reengineering and migration.

Gerardo Canfora received the Laurea degree in Electronic Engineering from the University of Naples "Federico II," Italy, in 1989. He is currently a full professor of Computer Science at the Faculty of Engineering and the Director of the Research Centre on Software Technology (RCOST) of the University of Sannio in Benevento, Italy. He has served on the program committees of a number of international conferences; he was program co-chair of the 1997 International Workshop on Program Comprehension and of the 2001 International Conference on Software Maintenance. His research interests include software maintenance, program comprehension, reverse engineering, reuse, reengineering, migration, workflow management, and document management.

Chih-Hung Chang received his BS and MS degree in Information Engineering and Computer Science from Feng-Chia University, Taiwan, in 1995 and 1997, respectively. He is currently a PhD student in Computer Engineering at Feng-Chia University, under the instruction of Dr. William C. Chu. His areas of research interest include software maintenance, software reengineering, design pattern, component-based software engineering, and reuse.

Ned Chapin is an information systems consultant with InfoSci Inc., USA. His decades of experience include all phases of the software life cycle and cover industrial, business, financial, non-profit, and governmental organizations. He has also served in roles from lecturer to professor of Information Systems at various universities. Ned's interests include software maintenance and evolution, database technology, systems analysis and design, and software management. He is a member of the ACM, ACH, AITP, IEEE Computer Society, and Sigma Xi—the Scientific Research Society of America. Ned is a Registered Professional Engineer, and a Certified Information Systems Auditor. His MBA is from the Graduate School of Business of the University of Chicago, and his PhD is from Illinois Institute of Technology. Ned currently is the co-editor of the *Journal of Software Maintenance and Evolution*. He can be contacted at: NedChapin@acm.org.

William C. Chu received his MS and PhD degrees from Northwestern University, Evanston, IL, in 1987 and 1989, respectively, all in Computer Science. Since 1998, he has been with the Department of Computer and Information Science at the Tunghai University, Taiwan, where he is now a professor and department head. From 1994 to 1998, he had been an associate professor at the Department of Information Engineering at the Feng Chia University, Taiwan. Prior to joining the Feng Chia University, he was a research scientist at Software Technology Center of Palo Alto Research Laboratories of Lockheed Missiles and Space Company, Inc., where he served as a member of Software Reverse Engineering and Software Reuse projects and a principal investigator of Converting Programming Languages to Ada project. He had received special contribution awards from Lockheed, both in the year of 1992 and 1993. In 1992, he was a Visiting Scholar to Department of Engineering Economic System at Standford University, where he was involved in projects related to the design and development of Intelligent Knowledge Based Expert System. His current research interests include the fields of software re engineering, maintenance, reuse, software quality, and E-commerce.

Yeh-Ching Chung received a BS degree in Computer Science from Chung Yuan Christian University in 1983, and the MS and PhD degrees in Computer and Information Science from Syracuse University in 1988 and 1992, respectively. He joined the Department of Information Engineering at Feng Chia University, Taiwan, as an associate professor in 1992. From 1998 to 2001, he was the chairman of the department. Since 1999, he is a professor in the department. His research interests include parallel compilers, parallel programming tools, mapping, scheduling, load balancing, cluster computing, and embedded systems. He is a member of the IEEE computer society and ACM.

Andrea De Lucia received the Laurea degree in Computer Science from the University of Salerno, Italy, in 1991, the MSc degree in Computer Science from the University of Durham, UK, in 1995, and the PhD degree in Electronic Engineering and Computer Science from the University of Naples "Federico II," Italy, in 1996. He is currently an associate professor of Computer Science at the Faculty of Engineering of the University of Sannio in Benevento, Italy. He serves on the program and organising committees of several international conferences and was program co-chair of the 2001 International Workshop on Program Comprehension. His research interests include software maintenance, reverse engineering, reuse, reengineering, migration, program comprehension, workflow management, document management, and visual languages.

Fabrizio Fioravanti obtained his MS and PhD from the University of Florence in 1996 and 2000, respectively. From 1992 he was a consultant for several companies covering different roles in several IT-related topics. During 2000 he was assigned professor of Computer Architecture for the BS in Electronic Engineering Program at the University of Florence. He was also co-chair and on the program committee for IEEE International Conferences. Since 2001, he is R&D Chief Officer for an Italian IT company. He has been a member of IEEE since 1992.

His research interests includes software engineering; object oriented metrics and methodologies; software metrics; and distributed systems.

Félix García has an MSc degree in Computer Science by the University of Castilla-La Mancha. He is an assistant professor at the Escuela Superior de Informática of the University of Castilla-La Mancha. He is a member of the Alarcos Research Group, in the same University, specialized in Information Systems, Databases and Software Engineering. Author of several papers related with software process modeling. His research interests are software processes and software measurement. He can be contacted at: Felix.Garcia@uclm.es

Hewijin Christine Jiau received a BE from National Cheng Kung University, Taiwan, an MS in EE/CS from Northwestern University, and a PhD in Computer Science from the University of Illinois at Urbana-Champaign. She is an assistant professor in the Department of Electrical Engineering, National Cheng Kung University, Taiwan. She was the senior engineer of the US Army CERL Laboratory. Her research interests include software engineering, object-oriented technology, database, and e-business.

Mira Kajko-Mattsson (PhD in Software Engineering) is a senior lecturer at Department of Computer and Systems Sciences (DSV), a joint department belonging both to Stockholm University and Royal Institute of Technology, Sweden. Kajko-Mattsson is the author of more than 20 international publications. Her present research area is software maintenance. Her early educational background lies in the humanities, which she had studied at the University of Poznan in Poland and Stockholm University in Sweden. In 1988, she received a BA degree in Data and Systems Sciences from Stockholm University. In 1998, she was awarded a PhL degree in Software Engineering. In 2001, she defended her PhD in Software Maintenance. Her email address is mira@dsv.su.se.

Chih-Wei Lu received his BS degree in Computer and Information Science from TungHai University in 1987 and his MS degree in Computer Science from University of Southern California in 1992. He is currently a PhD student in Computer Engineering at Feng-Chia University, Taiwan, under the instruction of Dr. William C. Chu. His areas of research interest include component-based software engineering, design pattern, software reuse, and software maintenance.

Mario Piattini, PhD and MSc in Computer Science (Madrid Technical University), MSc in Psychology (UNED), CISA (ISACA). He has been director of the development department of SiE and partner and founder of Cronos Iberica, S.A., where he has been director of the Training and Methodologies and Research and Development Departments. He has also worked as consultant to various organisations and companies like Ministerio de Industria y Energía, Ministerio de Administraciones Públicas, Ministerio de Interior, Siemens-Nixdorf, Unisys, Hewlett-Packard, Oracle, ICM, Atos-Ods, etc. He is an associate professor at the Escuela Superior de Informática of the Universidad de Castilla-La Mancha, Spain, at Ciudad Real, where he leads the Alarcos Research Group, specializing in Information Systems, Databases, and Software Engineering. He is the author of several books on these subjects, as well as a hundred papers in national and international conferences and magazines.

Macario Polo is a full-time professor at the Department of Computer Science in the University of Castilla-La Mancha, Spain. He has a PhD (University of Castilla-La Mancha) and a MS (University of Seville), both in Computer Science. He is author or co-author of several technical articles, papers, books, and book chapters, mainly related to software maintenance and software measurement. He has also published two novels in Spain. His e-mail address is Macario.Polo@uclm.es.

Bing Qiao having received his master's degree in Computer Science at Northwestern Polytechnical University in 2001, is working as a PhD student in the Software Technology Research Laboratory of De Montfort University, UK. His research interests include Web and Internet Software development and evolution.

Francisco Ruiz is a full time professor at the Department of Computer Science at University of Castilla-La Mancha (UCLM) in Ciudad Real, Spain. He was Dean of the Faculty of Computer Science for seven years until February 2000. Previously, he was Computer Services Director in the mentioned university and he has also worked in private companies as analyst-programmer and project manager of information systems. His current research interests include: software maintenance, software process technology and modeling, metrics for object-oriented databases, and methodologies for software projects planning and managing. In the past, other work topics have been: geographical information systems (GIS), educational software systems, and deductive databases. He has written twelve books on the previous topics and he has published fifty papers in national and international congresses, conferences, and journals. He belongs to various scientific and professional associations (ACM, IEEE-CS, ATI, AEC, AENOR, ISO JTC1/SC7, EASST, AENUI, and ACTA). His email address is Francisco.RuizG@uclm.es.

Norman F. Schneidewind is professor of Information Sciences and Director of the Software Metrics Lab at the Naval Postgraduate School, USA. Dr. Schneidewind is a Fellow of the IEEE, elected in 1992 for "contributions to software measurement models in reliability and metrics, and for leadership in advancing the field of software maintenance." In 2001, he received the IEEE "Reliability Engineer of the Year" award from the IEEE Reliability Society. The citation reads: "To Norman Schneidewind, in recognition of his contributions to Software Reliability Modeling and reliability leadership of key National purpose programs, such as the Space Shuttle Program." An article about the award was published in the *Monterey Herald*, 26 February 2002. An article about this award and other achievements appeared in the *U.S.C. Marshall School Magazine*, April 2002. In addition, Dr. Schneidewind was interviewed by National Public Radio about his work in software reliability engineering and distance learning in Montgomery, Alabama, on 1 April 2002.

Harry M. Sneed has a master's Degree in Information Sciences from the University of Maryland and has been active in the field of software maintenance since 1983, when he published a paper on regression testing for the first International Conference on Software Maintenance. Since then he has contributed continually to this ICSM conference of which he is in the program committee and to the *Journal of Software Evolution and Maintenance* of which he is on the editorial board. Mr. Sneed has also written three books on the subject of maintenance and developed more than 25 different maintenance tools for various programming languages from Assembler to Java. He is currently working as a quality control engineer and tool developer for a Viennese software house and as a part-time lecturer at the University of Regensburg, Bavaria.

Perdita Stevens is a lecturer and EPSRC Advanced Research Fellow in the Division of Informatics, University of Edinburgh. Her background includes working as a professional software engineer and a PhD in pure mathematics, and her research interests range from verification of infinite state systems to reengineering patterns, via UML tools. She has served on the programme committees of many international conferences, including the International Conference on Software Maintenance and the Working Conference on Reverse Engineering.

Hongji Yang is a reader in Computer Science at De Montfort University, Leicester, England. He leads the Software Evolution and Reengineering Group. His research interests include software engineering and distributed systems. He was program co-chair for IEEE International Conference on Software Maintenance (ICSM'1999), program co-chair for IEEE Future Trend in Distributed Computing Systems (FTDCS'2001) and program chair for IEEE Computer Software and Application Conference (COMPSAC'2002).

Index

H

I

L

M

O

P

Q

R

S

T

U

V

W

X